Kelly Maria Rego da Silva
Aldenora M. X. Rodrigues
Henrique B. Caminha

Practical Mycology:

Kelly Maria Rego da Silva
Aldenora M. X. Rodrigues
Henrique B. Caminha

Practical Mycology:

From concept to laboratory

Imprint
Any brand names and product names mentioned in this book are subject to trademark, brand or patent protection and are trademarks or registered trademarks of their respective holders. The use of brand names, product names, common names, trade names, product descriptions etc. even without a particular marking in this work is in no way to be construed to mean that such names may be regarded as unrestricted in respect of trademark and brand protection legislation and could thus be used by anyone.

Cover image: www.ingimage.com

This book is a translation from the original published under ISBN 978-613-9-76213-2.

Publisher:
Sciencia Scripts
is a trademark of
Dodo Books Indian Ocean Ltd. and OmniScriptum S.R.L publishing group

120 High Road, East Finchley, London, N2 9ED, United Kingdom
Str. Armeneasca 28/1, office 1, Chisinau MD-2012, Republic of Moldova, Europe
Printed at: see last page
ISBN: 978-620-6-42010-1

Summary

CHAPTER 1

FUNGI: FROM SAPROBIC TO PATHOGENIC

Henrique Barros Caminha

Fungi are ubiquitous organisms belonging to the *Fungi* Kingdom and are present in nature and all around us. When in nature, they act saprobically, degrading organic matter from plants and in the soil, acting as decomposers. They are present in our daily lives in various guises, and we are exposed to various fungal propagules on a daily basis, as their spores can be dispersed along with the wind, or be present in soil or water.

However, the vast majority of fungi do not act as disease-causing organisms in humans. They are important in industry, such as in the production of cheeses; beverages, such as wines and beers; and the production of inputs for maintaining health, such as certain antibiotics.

However, certain species can act as parasites of plants, animals and humans. In the case of humans, they are mainly responsible for opportunistic infections in immunocompromised individuals: those with Acquired Immunodeficiency Syndrome - AIDS, or transplant patients. On the other hand, certain fungal species are capable of infecting healthy humans, as a result of their greater virulence potential.

Fungal cell wall

The cell wall of fungi is a dynamic structure and is of paramount importance for cell viability, morphogenesis and pathogenesis. Its composition is closely related to the ecology of fungal species, and is subject to modification in response to environmental conditions and stress.

Certain constituents of the cell wall are conserved in different fungal species, while other components are species-specific, so this trade-off between variance and conservation makes the cell wall an organelle with high plasticity and phenotypic diversity. The turgor pressure, generated by the absorption of liquids and consequent rigidity of the fungal cell, applied through the cell membrane demonstrates how the cell wall is both malleable and robust.

This turgor pressure generates force that allows fungal hyphae to apply mechanical force to the substrates they are penetrating; this is a substantial factor in promoting parasitism and consequent infection in hosts. Furthermore, in relation to human pathogenic fungi, the cell wall induces innate and adaptive immune responses and, depending on its structure, can also act to evade the mechanisms of the host's immune system.

Carbohydrate polymers structure the elastic and robust core of the cell wall framework, where a variety of proteins and other surface constituents are added concomitantly to generate stability in the organelle. Overall, the basic framework is made up of an interconnected mesh of P-(1,3) glucan and chitin.

Most fungal cell walls are made up of overlapping layers, with the innermost layer showing a greater degree of structural conservation of the basic framework, unlike the outer layers, which are predominantly heterogeneous in their constituents and are related to the physiological specificities of each fungal species. The innermost layer of many fungal spores and so-called "black fungi" contains phenolic compounds, i.e. melanin, which provides protection against oxidising agents and exoenzymes.

This heterogeneity has great therapeutic value, as these constituents of the outermost layer will form certain virulence factors, such as adhesins responsible for attaching fungal structures to infected tissues, biofilms such as those seen in *Candida albicans*. Furthermore, certain polysaccharides that make up the outermost layer can be detected in biological fluids, thus allowing detection by serological tests in infected individuals.

Galactomannan, mannan and glucuroncxylomannan represent specific cell wall polysaccharides of interest for the detection of aspergillosis, candidiasis and cryptococcosis, respectively.

Fungal morphology

Yeast-like fungi are those whose primary morphological form is a unicellular form that divides by budding or binary fission, i.e. yeast. In budding, the original cell (mother cell) will give rise to an adjacent cell (daughter cell) which will detach and give rise to a new yeast. The vegetative growth unit of filamentous fungi is the hypha, which may or may not have septa (divisions) or adjacent branches. When there is melanin in their cell wall, the hyphae are called "dematiaceous hyphae", as in the case of fungi that cause chromoblastomycosis, for example, *Exophiala spinifera*; when there is no pigmentation, they are called hyaline hyphae, as in *Aspergillus flavus*.

Fungal reproduction

The nucleus of the fungal cell can be compacted to form asexual spores, or fuse to give rise to sexual spores after meiosis. In most fungi, sexual spores are considered to be more resistant to environmental weathering than asexual spores. Certain sexual structures, such as basidiocarps, promote the dispersal of spores by wind, water or animals. Sexual reproduction provides for the succession of haploid and diploid phases, transitions carried out during meiosis, where recombination and chromosomal segregation and syngamy occur with the consequent fusion of two haploid cells.

Most fungi are capable of both asexual and sexual life cycles and have mechanisms that strictly control the *mating* process, such as temporal control, gamete dispersal, cell and nuclear recognition and fusion. In relation to the *mating* process, regions in specific *loci of* the fungal genome will designate the *mating-types,* i.e. phenotypic characteristics that will influence the recognition and subsequent fusion of fungal cells (syngamy). In homothallic fungi, syngamy can occur in both the same and different *mating-types.* For heterothallic fungi to reproduce successfully, the individuals in a given fungal population must have different *mating-types,* as specimens with the same *mating-type* are incompatible for the production of sexual spores.

Once cell recognition has taken place, syngamy occurs, generating a dikaryotic state; then meiosis occurs with the consequent establishment of the diploid state and, as a result of meiotic divisions, haploid sexual spores are generated. As fungi do not have typical sexual organs, *mating-types* and conditions of homo- or heterothallism act to control sexual reproduction. Furthermore, the sexual state can also be referred to as teleomorph or perfect; the asexual state as anamorph or imperfect.

Spores can also be generated asexually, by cell transformation, a complex mechanism which will result in the generation of specialised structures; such as the modification and specialisation of hyphae into conidia of *Aspergillus nidulans.*

Spores are given different names depending on the fungal class and its generation, i.e. sexual or asexual. In addition, spores can be packaged in structures called fruiting bodies, which vary in size and shape between different fungal species.

Fungal dimorphism

Certain species of fungi can vary their morphology, alternating between hyphae and yeasts, and vice versa. Such as the dimorphism observed in growth according to temperature in the case of the ascomycetes *Histoplasma capsulatum* , *Blastomyces dermatitidis, Talaromyces marneffei* (*Penicillium marneffei), Coccidioides immitis, Paracoccidioides brasiliensis* and *Sporothrix schenckii* ; all have a filamentous form when grown at 25 C° and a yeast-like form when grown at 37 C°.

All of the aforementioned ascomycetes are found saprobically in the environment, with the yeast-like form being found in the tissues of infected individuals; this illustrates the relationship between dimorphism as a specialised strategy for colonising the host. Another example is the dimorphism observed in the yeast *Candida albicans,* when parasitised it has the form of pseudohyphae and true hyphae. The pseudohypha is made up of elongated budding cells, connected to each other and with a visible constriction in the budding region, thus mimicking a hypha. The true hypha of *C. albicans*, on the other hand, has parallel sides with no constriction of the cell wall.

REFERENCES

AWAN, A. R. et al. Biosynthesis of the antibiotic nonribosomal peptide penicillin in baker's yeast. **Nature Communications**, 2017.v. 8, n. 15202.

BENNETT, R. J.; TURGEON, B. G. Fungal Sex: The Ascomycota. In: **The Fungal Kingdom**. [s.l.] American Society of Microbiology, 2016. v. 4p. 117-145.

BILLIARD, S. et al. Sex, outcrossing and mating types: unsolved questions in fungi and beyond. **Journal of Evolutionary Biology**, v. 25, n. 6, p. 1020-1038, jun. 2012.

BONHOMME, J.; D'ENFERT, C. Candida albicans biofilms: building a heterogeneous, drug-tolerant environment. **Current Opinion in Microbiology**, v. 16, n. 4, p. 398-403, Aug. 2013.

BOWEN, A. et al. A Monoclonal Antibody to Cryptococcus neoformans Glucuronoxylomannan Manifests Hydrolytic Activity for Both Peptides and Polysaccharides. **Journal of Biological Chemistry**, v. 292, n. 2, p. 417-434, 13 Jan. 2017.

CORONADO, J. E. et al. Conserved processes and lineage-specific proteins in fungal cell wall evolution. **Eukaryotic cell**, v. 6, n. 12, p. 2269-77, dec. 2007.

DABOIT, T. C. et al. A case of Exophiala spinifera infection in Southern Brazil: Molecular identification and antifungal susceptibility. **Medical Mycology Case Reports**, v. 1, n. 1, p. 72-75, 2012.

DIJKSTERHUIS, J. The fungal spore and food spoilage. **Current Opinion in Food Science**, v. 17, p. 68-74, 2017.

DIJKSTERHUIS, J. Fungal spores: Highly variable and stress-resistant vehicles for distribution and spoilage. **Food Microbiology**, Nov. 2018.

ERWIG, L. P.; GOW, N. A. R. Interactions of fungal pathogens with phagocytes. **Nature Reviews Microbiology**, v. 14, n. 3, p. 163-176, 8 Mar. 2016.

GARCIA-GIRALDO, A. M. et al. Invasive fungal infection by Aspergillus flavus in immunocompetent hosts: A case series and literature review. **Medical mycology case reports**, v. 23, p. 12-15, mar. 2019.

HELBING, U. New developments for mixing and metering. **Chemical Fibers International**, v. 57, n. 4, p. 188-192, 2007.

KHAN, A.; EL-CHARABATY, E.; EL-SAYEGH, S. Fungal infections in renal transplant patients. **Journal of clinical medicine research**, v. 7, n. 6, p. 371-8, jun. 2015.

KNOP, M. Yeast cell morphology and sexual reproduction - A short overview and some considerations. **Comptes Rendus Biologies**, v. 334, n 8-9, p. 599-606, Aug. 2011.

KÚES, U. From two to many: Multiple mating types in Basidiomycetes. **Fungal Biology Reviews**, v. 29, n. 3-4, p. 126-166, dec. 2015.

LATGÉ, J.-P. The cell wall: a carbohydrate armour for the fungal cell. **Molecular Microbiology**, v. 66, n. 2, p. 279-290, oct. 2007.

LIMPER, A. H. et al. Fungal infections in HIV/AIDS. **The Lancet. Infectious diseases**, v. 17, n. 11, p. e334-e343, 2017.

MONEY, N. P. Insights on the mechanics of hyphal growth. **Fungal Biology Reviews**, v. 22, n. 2, p. 71-76, May 2008.

NIELSEN, K.; HEITMAN, J. Sex and Virulence of Human Pathogenic Fungi. In: **Advances in Genetics**. [s.l: s.n.]. v. 57p. 143-173.

PERCIVAL, A.; THORKILDSON, P.; KOZEL, T. R. Monoclonal Antibodies Specific for Immunorecessive Epitopes of Glucuronoxylomannan, the Major Capsular Polysaccharide of Cryptococcus neoformans, Reduce Serotype Bias in an Immunoassay for Cryptococcal Antigen. **Clinical and Vaccine Immunology**, v. 18, n. 8, p. 12921296, aug. 2011.

RAUDASKOSKI, M. Mating-type genes and hyphal fusions in filamentous basidiomycetes. **Fungal Biology Reviews**, v. 29, n. 3-4, p. 179-193, 2015.

RIQUELME, M. et al. Fungal Morphogenesis, from the Polarised Growth of Hyphae to Complex Reproduction and Infection Structures. **Microbiology and Molecular Biology Reviews**, v. 82, n. 2, p. 1-47, 2018.

STEINBERG, G.; SCHUSTER, M. The dynamic fungal cell. **Fungal Biology Reviews**, v. 25, n. 1, p. 14-37, mar. 2011.

YU, J.-H. Regulation of Development in *Aspergillus nidulans* and *Aspergillus fumigatus*. **Mycobiology**, v. 38, n. 4, p. 229, 2010.

CHAPTER 2

LABORATORY DIAGNOSIS OF YEASTS

Brenda de Araujo Barbosa
Francisco Anderson Fortuna de Carvalho Teixeira Hernande
Pereira Passos Junior José Mario Lima Coutinho Tainá Vieira
Sousa Lopes

Identification of yeasts

Morphological

For a long time fungi were considered to be plants, but only since 1969 have they been classified in a separate kingdom: *Kingdom Fungi*. They are determined by their characteristics, which differ from those of the plant kingdom, such as:

❖ They do not synthesise chlorophyll;

❖ They have a rigid wall that can be composed of cellulose, glycans, mannans or chitin in their cell membrane, with sterols present;

❖ Some aquatic fungi have cellulose;

❖ It doesn't store starch as a reserve, its main material is glycogen;

Within the Fungi kingdom there are further subdivisions such as the definition of filamentous fungi (moulds and mildews) and yeasts, which are eukaryotic organisms, pluricellular and unicellular respectively, heterotrophic, and generally found in humid habitats rich in organic matter.

Yeasts are unicellular, non-filamentous, have an average diameter of 1 to 5 um and a length of 5 to 30 |im. The yeast cells have an oval or spherical shape and can be elongated, they have no flagella and are immobile, reproducing asexually by budding. These are fungi from the class of ascomycetes, which belong to the phylum Ascomycota. This phylum is characterised by fungi in which spore production takes place in specific sporangerangia, called asci.

They develop through alcoholic fermentation and have a well-defined cell membrane, not very thick in young cells, but very rigid in adults. They are chemo- heterotrophic and facultative aerobic (oxidative and fermentative metabolisms).

While they can reproduce asexually and sexually, they are often unable to utilise starch and cellulose as a carbon source.

The size of yeast cells varies depending on the culture. Young yeast cells can be very uniform in some species or extremely heterogeneous in   others. This disparity is often used to differentiate between species and strains of the same species, but in general they vary considerably in terms of their size.

Yeasts can have various forms, which can be the result of the way they reproduce, the environment, humidity or good cultivation conditions and the age of the culture, as represented below by their forms and the agents that determine their characterisation.

The structure of the yeast cell and information on its cytology has been obtained through direct observations under an optical microscope, staining techniques, transmission electron microscopy and scanning microscopy. Thus, it is possible to see that the main structures of yeasts are;

Forma		Principais Representantes
	Redondas	*Trigonopsis* *Candida* *Saccharomyces*
	Ovais	*Hansenula* *Saccharomyces* *Trigonopsis*
	Cilíndricas	*Hansenula* *Saccharomyces* *Kloëckera*
	Triangulares	*Trigonopsis*
	Apiculares	*Kloëckera*
	Ogivas	*Bretanomyces*

Source: CTISM

❖ **Cell wall:** is responsible for the shape of yeast. The wall is made up of glycan (30 to 34 per cent) and mannan (30 per cent). It is thin in young cells and thick in adult cells. Proteins are also present in yeast cell walls, around 6 to 8 per cent; lipids vary from 8.5 to 13.5 per cent. The amount of chitin varies according to the species.

❖ **Cytoplasmic membrane:** located below the cell wall, its function is to allow the selective entry of nutrients and protect the yeast from the loss of small molecules due to leakage from the cytoplasm. The chemical composition of the membrane is made up of glycoproteins, lipids and ergosterol (unlike mammalian membranes which contain cholesterol).

❖ **Surface structures:** some yeasts are covered in a slimy, viscous and adherent material, which is the capsular substance. Most yeast capsules are made up of polysaccharide.

❖ **The cytoplasm: this** is the fluid in which the organelles are located. It contains enzymes, reserve carbohydrates (glycogen and trehalose) and large quantities of ribosomes and polyphosphate. Between 1 and 5 per cent of yeast DNA may be present in the cytoplasm.

❖ **The cell nucleus:** easily seen in the light field of an optical microscope. It is well-defined, small, spherical or reniform, surrounded by a semi-permeable membrane and has metabolic and reproductive functions.

❖ **Vacuoles:** when yeast cells are viewed under a phase contrast microscope, one or more vacuoles of different sizes (0.3 to 3.0 mm in diameter) can be seen. They have a spherical appearance and are more transparent to a beam of light than the surrounding cytoplasm. Under an electron microscope it can be seen that the vacuole is surrounded by a simple membrane and its constitution is related to the transport of substances that are stored in the vacuole, such as enzymes, free amino acids and lipids. Vacuoles also serve as storage vesicles for various hydrolytic enzymes.

❖ **Mitochondria:** these are membranous organelles that are usually close to the periphery of the cell, but in aerobic yeasts they are distributed throughout the cytoplasm. The number of mitochondria can vary from one to twenty per cell. These organelles are surrounded by outer and inner membranes; the inner membrane forms the mitochondrial cristae. They are important in aerobic energy conversion processes.

BIOCHEMISTRY

In order to isolate and identify the main yeast species, the mycology laboratory must carry out certain biochemical tests and physiological tests. The most common of these are simple and used to identify Candida species such as *Candida albicans and Candida sp. They* are: slide culture to observe filaments and germ tubes.

In slide culture, the first to be carried out, the production of branched hyaline hyphae that can fragment into spores, called arthroconidia, is evaluated. The genera *Geotrichum and Trichosporon* can present this type of hyphae. The presence or absence of fragmentation in the hyphae is an important point to observe under microscopy, and a species of the Candida genus can be identified by the formation of chlamydospores, which is characteristic of *Candida albicans.*

In addition to searching for the germ tube, which differentiates the C. albicans species from the others in the Candida genus, there are also other genera that can be identified by specific characteristics, such as the *Cryptococcus genus,* which is searched for a polysaccharide capsule that gives the fungus protection.

Cultivation is done directly on a slide, where a drop of India ink and a section of the culture can be used to visualise the capsule, which appears as a clear halo around the *Cryptococcus* blastoconidia and contrasts with the black background of the slide. However, the urease test is still available in the laboratory, which allows the identification of fungi that have a salmon or reddish colour on their macroscopy, such as the genera **Cryptococcus** *sp.* and **Rhodotorula.**

If these tests don't identify the species or even the genus of the fungus being studied, there are some tests that use carbohydrate assimilation, also known as Zymograms, or those that use carbon and nitrogen, called auxanograms.

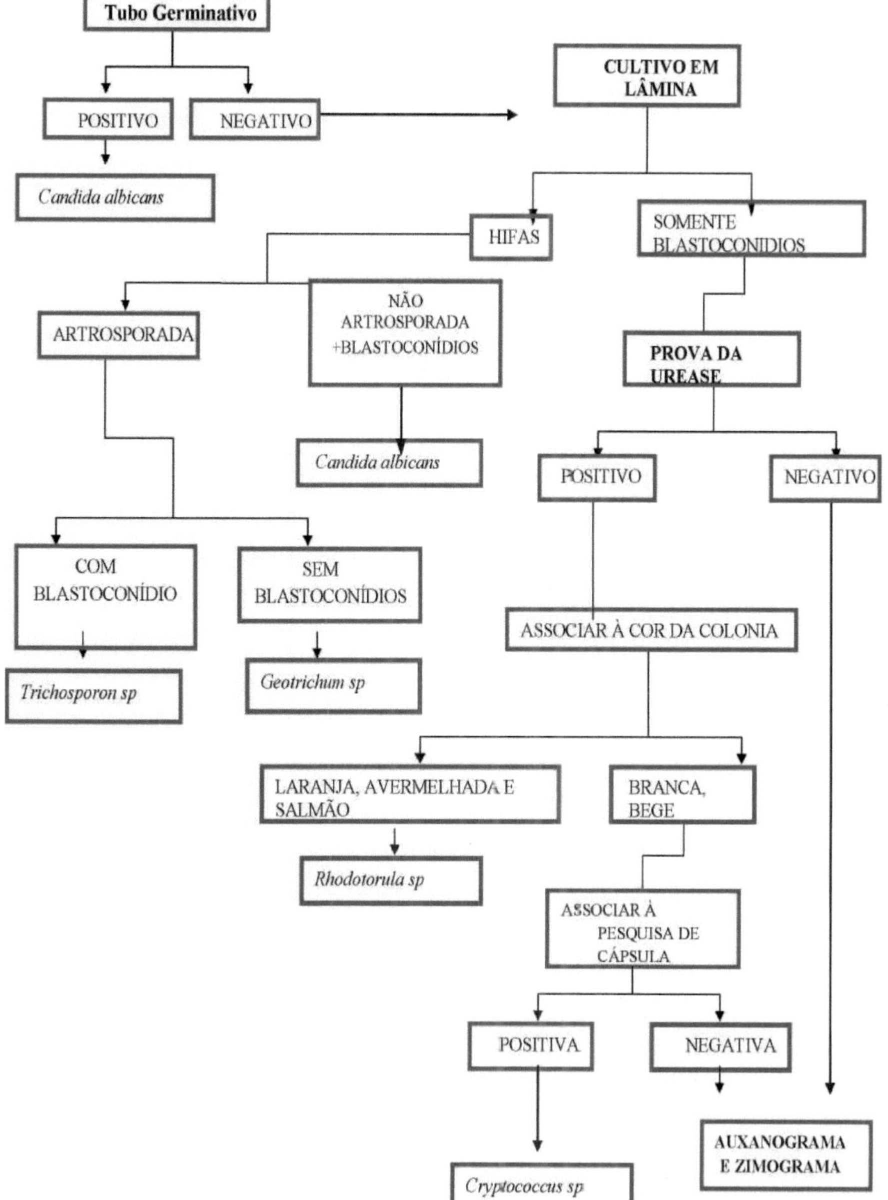

Simplified scheme for identifying some yeast genera:

GERM TUBE TEST

The germ tube identification technique is used to identify *Candida* albicans species. It consists of identifying the germ tube due to the species' ability to form them earlier than other forms of *Candida*, which facilitates their identification and differentiation.

The growth and differentiation of the yeast form into a germ tube is regulated by a number of factors, such as temperature, pH and media containing substances that will lead to the formation process (serum, N-acetyl-D-glucosamine, L-proline, ethanol) and the presence of chelating agents that will inactivate disulfhydryl reductase, which is responsible for reducing the disulfide bridges in the cell wall.

Its methodology involves inoculating the yeast into human, rabbit, foetal bovine or horse serum. It stimulates the yeast to emerge its germ tube when it comes into contact with the serum, thus making it possible to identify Candida albicans (germ tube present) or other yeast species (germ tube absent).

> **Procedure**

❖ Using a platinum loop or Pasteur pipette, a small aliquot is removed from the yeast colony.

❖ Mix it in a tube containing human, rabbit, foetal bovine or horse serum.

❖ The tube containing the suspension is incubated at 37°C for 2 to 3 hours

❖ At the end of incubation, a small sample of the suspension is taken for microscopic analysis.

When analysing the sample, it is possible to observe the presence or absence of a germ tube; when present, they have no septum and a constriction at the inclusion point between their elongation and the blastoconidium. This differs from pseudohyphae which may or may not be septate, but which have a constriction point.

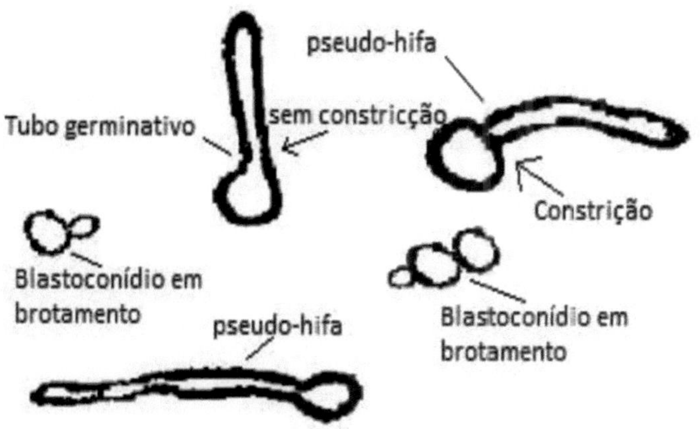

Source: http://www.ebah.com.br/content/ABAAAA8XAAA/daty-deteccao-identificacao-fungos-anvisa?part=4.

SLIDE CULTURE FOR FILAMENTATION AND CHLAMYDOSPORE TESTS

Slide cultivation is carried out after the germ tube test is negative for some types of yeast found in the Candida genus. It is one of the most commonly used physiological tests for identifying Candida *albicans* and *Candida sp,* which occurs by identifying the production of branched hyaline hyphae that can fragment into spores, known as arthroconidia.

Arthroconidia are often present in the genera *Geotrichum* and *Trichosporon*. During the slide culture test, yeasts that may belong to the genus Candida will form branched hyaline hyphae without fragmentation and yeasts present in the genus *Candida Albicans* commonly form chlamydospores, characteristic of this genus.

> **Method:**

3 ml of molten cornmeal agar is added to a sterilised slide, which should be placed on a glass rod or another slide used as a support in the Petri dish, then the yeast is sown by making two parallel grooves with an "L" shaped needle, then covered with a sterile coverslip.

During the process, the plate should always be moist to favour cultivation. To maintain humidity, the plate should be made into a humid chamber by adding 2 ml of sterile distilled water or a moistened cotton pad inside the plate, thus preventing the sowing from becoming dry.

The plate remains in culture for 24 to 72 hours, then the preparation is examined under an optical microscope. If there are characteristics of yeasts of the genus Candida, the presence of hyaline, septate and branched hyphae is noted, but if chlamydospores are present, it is presumptive of *Candida Albicans.*

Geotrichum have a unique form of arthrospores that allow them to be identified quickly, but if blastoconidia are present, it is unique to *Trichosporon.* The medium for cultivation can be found on the market, but it is necessary to add surfactants called (Tween 80) to improve the effectiveness of the method.

Agar-cornmeal medium:

❖ Cornmeal 20 g

❖ Agar 10 g

❖ Tween 80 5 ml

> **Preparation:**

❖ Boil 250 ml of water to boiling point and add the cornmeal;

❖ The cornmeal should be filtered through gauze folded in four;

❖ Add the Agar to 250 ml of water;

❖ Add hot water to bring the volume of the infusion back to 250 ml;

❖ Add the suspension;

❖ pH 6.6-6.8, adjusting if necessary;

❖ Add Tween 80;

❖ Distribute 5 ml volumes into tubes and sterilise in an autoclave at 120°C for 15 minutes;

UREASE TEST

It is a technique for identifying yeast-like fungi that makes it possible to identify the genus *Cryptococcus sp,* when positive and with morphological analysis of the colony. It is widely used in mycology laboratories to identify yeasts of the *Rhodotorula genus.* To differentiate between the two species, which are urease positive, the colour of the colony is observed, with *Cryptococcus* having a reddish or salmon pigmentation.

The methodology consists of sowing yeasts on the surface of the urease medium (Christensen's urea agar) and then incubating them, analysing whether there is a change in the medium to a bishop's pink colour within 24 hours, indicating that the colony is positive.

> Procedure

- ❖ Using a platinum loop, sow a batch of yeast onto the surface of the urease medium;
- ❖ After sowing, incubate for 24 hours;
- ❖ Identify if there is a change in colour;

When analysing the change in colour (positive urease test), it is important to analyse the colony due to the ability of other fungal genera to produce urease, taking into account their morphological aspects. If the sample is negative, the auxanogram and zymogram tests are used to identify it.

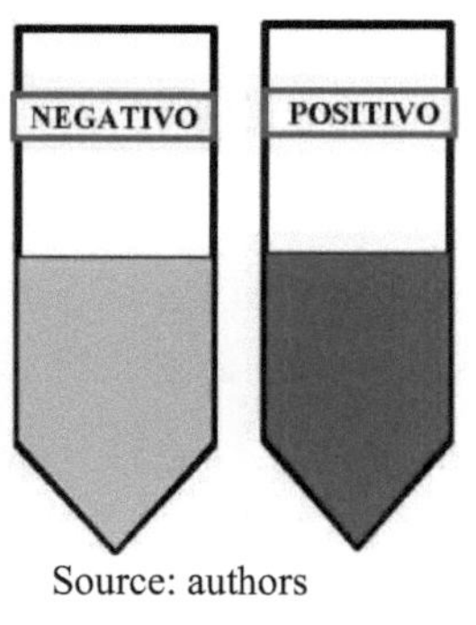

Source: authors

AUXANOGRAM OR ASSIMILATION TEST FOR CARBON AND NITROGEN SOURCES

The identification of yeasts is fundamentally based on the carbohydrate and nitrogen assimilation tests as well as the carbohydrate fermentation test.

It is the ability of a yeast to grow aerobically in the presence of a certain carbohydrate supplied as the sole source of energy. It can be carried out in liquid or solid basal medium, the latter being more suited to routine use because it is simpler to carry out and provides earlier and more accurate readings. The auxanographic technique or assimilation on solid media uses agar media devoid of any carbon source, added to the yeast suspension and distributed on a plate. After solidification, the carbohydrates are aliquoted onto the agar.

Filter paper discs impregnated with different carbohydrates can be used, either commercially available or made in the laboratory as an alternative. Positivity is assessed by observing a halo of yeast growth in the presence of the carbohydrate provided. The following can be used as carbon sources: dextrose, maltose, sucrose, galactose, lactose, trehalose, melibiose, L-arabinose, cellobiose, xylose, raffinose, dulcitol, rhamnose, inulin, mannitol, inositol and others.

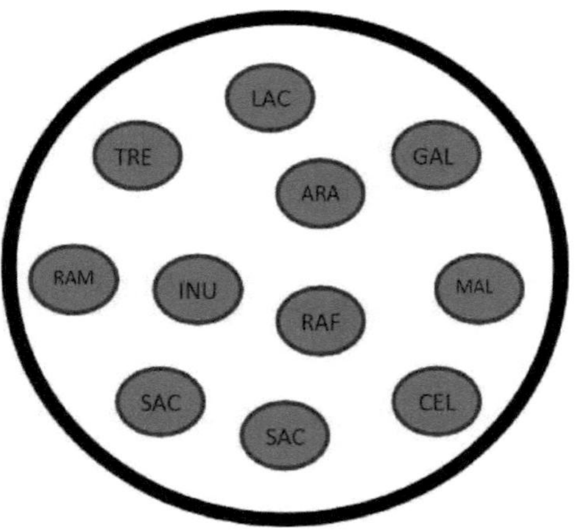

Auxanogram (diagram) - sugar assimilation test. 11 carbohydrates used: glucose, sucrose, trehalose, maltose, galactose, cellobiose, inulin, lactose, arabinose, raffinose, rhamnose. Observe sugar assimilation in GLI (glucose), GAL (galactose), MAL (maltose), SAC (sucrose) and TRE (trehalose).

It consists of a yeast's ability to grow aerobically in the presence of a nitrogenous compound supplied as the sole source of energy. As with the carbon assimilation test, this test can also be carried out in liquid or solid basal medium, where each laboratory standardises one of the techniques that best suits its routine.

The assimilation of nitrogen sources in solid media uses agar media devoid of any nitrogen source, added to the yeast suspension and distributed on a plate. After solidification, nitrogen compounds are aliquoted onto the agar. Positivity is assessed by the growth of the yeast in the presence of the nitrogen compound supplied. Peptone (positive control) and potassium nitrate are generally used as nitrogen sources, but other compounds can be used.

ZYMOGRAM OR FERMENTATION OF SUGARS

Carbohydrate fermentation is the ability of yeast to grow anaerobically in the presence of a certain sugar provided as the sole source of energy, observed through the production of carbon dioxide and a change in pH. It can be carried out in liquid or semi-solid media.

It is carried out by inoculating a yeast suspension into a test tube containing culture medium, a 2% solution of the desired sugar and an inverted Durham tube. Positivity is observed through the production of gas inside the Durham tube and the change in colour of the medium when a pH indicator is used. With few exceptions, it is observed that when a yeast uses a carbohydrate source fermentatively, it is also capable of using it oxidatively, but the opposite is not true. The carbohydrates dextrose, maltose, sucrose, galactose, lactose and trehalose are usually used for the test, but others can be used if necessary

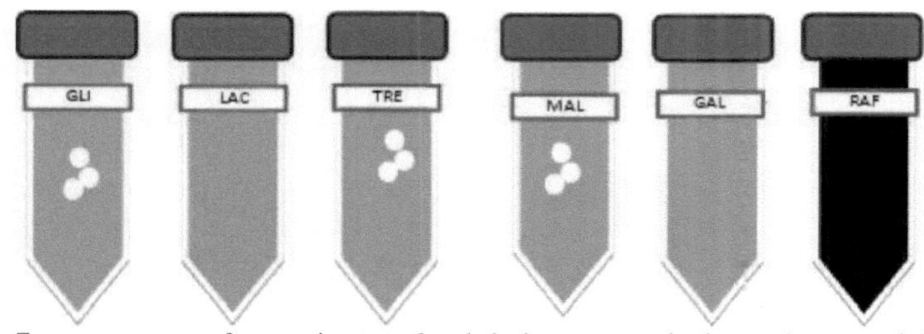

Zymogram - sugar fermentation test. 6 carbohydrates are used: glucose, lactose, trehalose, maltose, galactose and raffinose. Note gas formation in GLI (glucose), TRE (trehalose) and MAL (maltose). Source: authors

IDENTIFICATION OF YEASTS OF MEDICAL INTEREST

Fungi are eukaryotic microorganisms that produce spores, neither produce nor synthesise chlorophyll, absorb nutrients and have asexual reproduction by budding or even sexual reproduction. These organisms are ubiquitous, i.e. found everywhere, be it soil, water, air, crops, etc.

Mycoses are diseases caused by pathogenic fungi, which can cause superficial, opportunistic, deep benign or malignant clinical conditions and can lead to death. Over the years, fungal diseases have gained substantial attention, with the advent of sexually transmitted diseases (STDs), advances in basic disease therapies, the indiscriminate use of antibiotics (Antibiotic Therapy), improved transplant techniques, in other words, greater survival of patients with various illnesses.

Candida sp species reside as commensals, being residents of the normal microbiota of healthy individuals. However, when there is an imbalance in the normal flora or when the immune system is compromised, species of the *Candida* genus tend to proliferate, becoming pathogenic. Candidiasis, a set of clinical syndromes caused by this genus of fungus, can have endogenous or exogenous causes. Endogenous causes come from the normal microbiota and exogenous causes from the external environment, such as STDs.

There are several species of Candida, among which Candida *albicans* is the most infectious and most commonly found in clinically isolated samples, around 60 %. Other species can also be infectious, but less frequently, such as Cancida *glabrata, Candida parapsilosis, Candida tropicalis, Candida krusei, Candida lusitaniae* and Candida *guilliermondii*. The *C. glabrata* and *C. krusei* species are resistant to the antifungal Flucorazole, and their occurrence is related to the increased use of this drug in hospitals.

Candida albicans is a dimorphic fungus, i.e. depending on the temperature it can have a yeast-like or filamentous form (pseudohyphae and true hyphae), observed in pathogenic processes. In addition, under optimum growth conditions, the fungus can produce chlamydospores, i.e. rounded spores with a thick cell wall.

Diagnostic flowchart for the Candida albicans species.

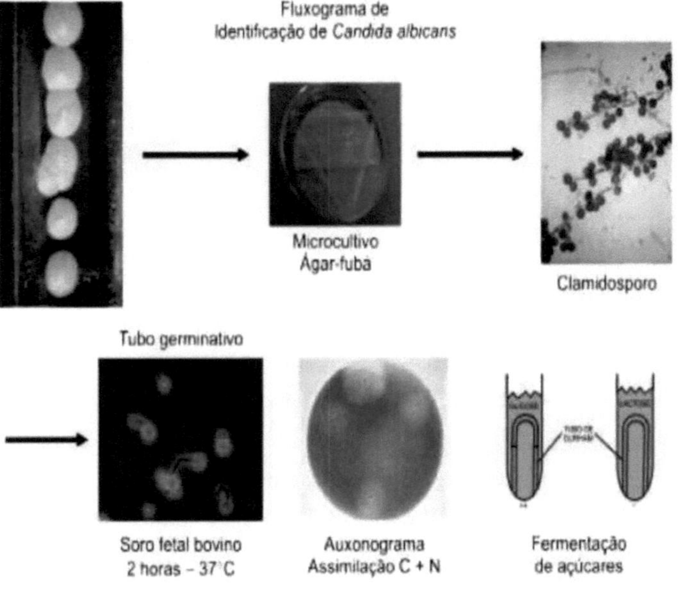

The genus *Trichosporon* causes "White Piedra", a superficial fungal infection that affects the coat, also known as Tricosporonosis. This is characterised by the formation of irregular nodules on the hair that are visible to the naked eye. However, it is common in patients with compromised and weakened immune systems and has a high morbidity and mortality rate.

This genus can be classified into several species, *including T. ovoides, T. inkin, T. asahii, T. asteroides, T. mucoides and T. cutaneum (or T. beigelli).* These fungal species can cause infections in both immunocompromised and healthy patients.

The main species of this genus, *Trichosporon beigelli,* presents the invasive form, especially in debilitated and hospitalised patients. The *Trichosporon* genus is a yeast-like fungus that develops rapidly in culture media, forming yellowish-white colonies, similar to those of the Candida genus, initially smooth and over time becoming rough.

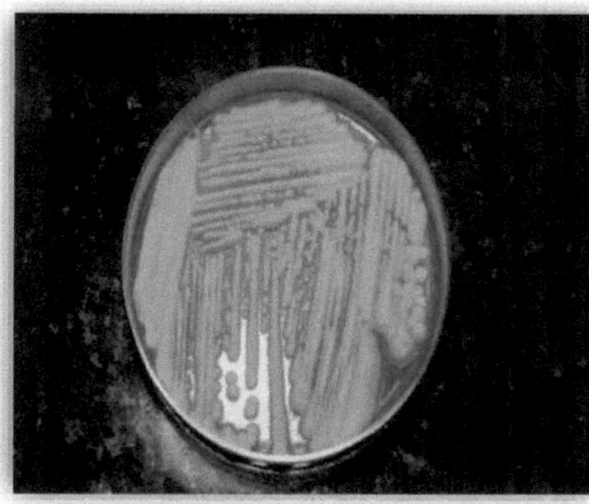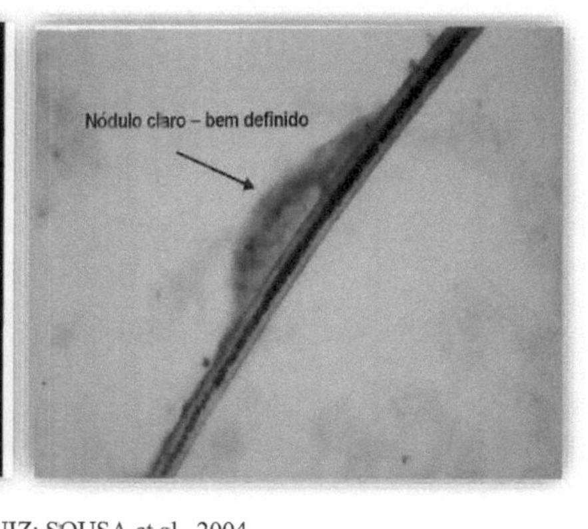

Source: ZAITZ; RUIZ; SOUSA et al., 2004.

Currently described species of medical interest:

1-Trichosporon beigelii: Causative agent of "Piedra branca" on hairy surface areas.	**5-Trichosporon ovoides**: Agent related to superficial infections.
2-Trichosporon inkin: Agent involved in "Piedra branca" in the genital area.	**6-Trichosporon asahii**: Agent involved in deep mycoses.
3-Trichosporon asteroides: Agent of "Shallow white stone.	**7-Trichosporon mucoides**: Agent of deep mycoses in children and neonates.
4-Trichosporon pullulans: Rarely causes systemic infection agent of mycosis of the oral cavity in immunocompromised individuals.	

Source GUEHO, 1994 revised by SUGITA, 1998

The *Geotrichum fungus* is commonly isolated from soil, water, air and plants, as well as in bovine production, and can be found in the normal flora of humans. This genus includes severe species such as *Geotrichum candidum. Geotrichum clavatum* and *Geotrichum fici.* The Geotrichum *candidum* species is a widely distributed fungus and is the causative agent of the disease known as geotrichosis, which affects the respiratory system, particularly the bronchopulmonary, oral, vaginal and cutaneous cavities.

In addition, it can appear in the tissues, which is considered a true diagnosis of geotrichosis. *Geotrichum fici* has a *strong*, intense odour reminiscent of pineapple. The teleomorphic form shows hyaline, subglobose hyphae. It has similar characteristics to the genus *Tricosporon sp,* differing in the presence of arthroconidia. The genus *Geotrichum sp.* has white, septate, unicellular mycelium and no conidiophores. It has short, cylindrical ends and can form rectangular basidiomycetes.

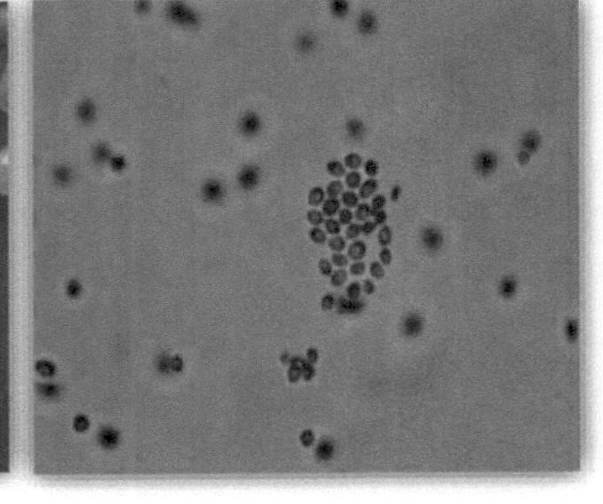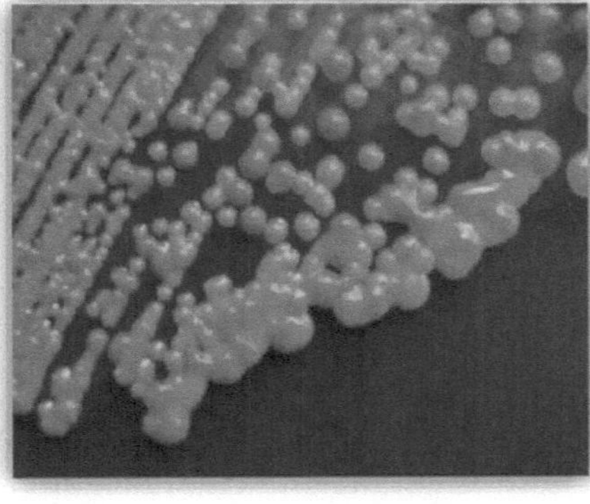

Source: NUPEMAS-FMVZ, UNESP, Botucatu, São Paulo.
Paulo.

Rhodotorula sp. belongs to the class Basidiomycota, is a common contaminant in the laboratory and can cause systemic infections of the lungs, kidneys and central nervous system (CNS), and is associated with hospital-acquired infections caused by contaminated venous catheters, prostheses and kidney valves.

Its colonies are fast-growing, coral to pink in colour and can be orange or red, mucoid in appearance, soft, smooth and moist.
To differentiate the species of the genus Rhodotorula, carbohydrate assimilation is used, specifically maltose. R mucilaginosa sometimes assimilates, R.glutinis always assimilates and R, minuta never assimilates. The latter sometimes assimilates lactose.

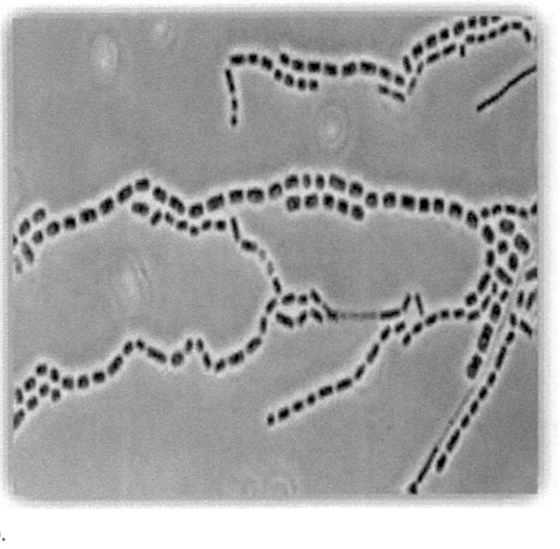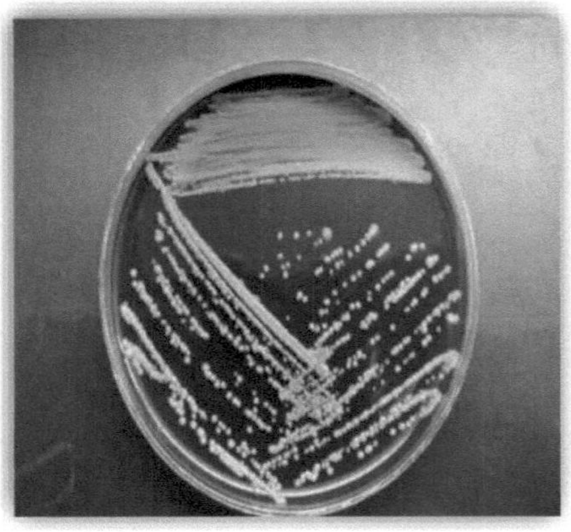

Source: Bonifaz, 2012.

Yeast	T g	Cultivation/Lamina		Ur	Assimilation								Fermentation				
		hifa	Air		Sa	Ma	La	Ce	Tr	Ra	X	I	T r	G I	S A	M a	R a
C. albicans	+	+	-	-	+	+	-	-	+	+	+	-	V	+	-	+	-
C. tropicalis	-	+	-	-	+	+		V	+	+	+	-	+	+	V	+	-
C. parapsilosis	-	+	-	-	+	+	-	V	+	+	+	-	V	+	-	-	-
C. krusei	-	+	-	-	-	-	-	-	-	-	-	-	-	+	-	-	-
C.guilliermondii	-	+	-	-	+	+	-	+	+	+	+	-	V	+	+	-	-
C. glabrata	-	-	-	-	-	-	-	-	+	-	-	-	+	+	-	-	-
C. neoformans	-	-	-	+	+	+	-	V	+	V	+	+	-	-	-	-	-
Geotrichum	-	+	+	-	-	-	-	-		-	+	-	-	V	-	-	-
Trichosporon	-	+	+	V	+	+	+	+	V	V	+	V	-	-	-	-	-
Rhodotorula sp	-	-	-	+	+	V	^B	V	+	+	+	-	-	-	^B	-	-
Saccharomyces	-	-	-	-	+	+	-	-	V	+	-	-	V	+	+	+	-

Tg = germ tube, Ar = arthrospore, Ur = urease, Sa = sucrose, Ma = maltose, La = lactose, Ce = cellubiose, Tr = trehalose, Ra = raffinose, X = xylose, I = inositol, NO3 = nitrate, Gl = glucose, + = pos, - = neg, V = variable.

REFERENCES

NATIONAL HEALTH SURVEILLANCE AGENCY. Manual of Clinical Microbiology for Infection Control in Health Services. Salvador: ANVISA, 2004. Irregular pagination.

ALBERTS, B. et al. Fundamentals of Cell Biology. 2. ed. Porto Alegre: Artmed, 2006.

AMABIS, J. M.; MARTHO, G. R. Biologia. 2ª série. V. 2. São Paulo: Moderna, 2004

AMORIM, H.V.; BASSO, L.C.; OLIVEIRA, A.J. de. Study of contaminating yeasts at the Santo Alexandre Mill. In: ESCOLA SUPERIOR DE AGRICULTURA "LUIZ DE

ARAUJO, I.R.M.; MARQUES, S.G. Trichosporonosis: etiology, clinical manifestations, laboratory diagnosis, epidemiology and treatment. São Luís, MA.

BARBEDO, L.S.; SGARBI, D.B.G. Candidiasis. Brazilian Journal of Sexually Transmitted Diseases, Rio de Janeiro. 2010; 22(1):22-38.

BARNETT, H.L. & BARRY B. H. IIIllustrated Genera of Imperfect Fungi. 4 ed. 1998.

CHAFFIN, W.L. *et al.* Cell wall and secreted proteins of *Candida albicans:* identification, function, and expression. *Microbiol Molec Biol Rev*, v. 62, p. 130-80, 1998.

GENTE, S; DESMASURES, N; JACOPIN, C; et al. Food Microbiology. June 2002.

JUNQUEIRA, L. C.; CARNEIRO, J. Biologia Celular e Molecular. Rio de Janeiro: Guanabara Koogan, 2005.

KAUFFMAN, C.A. Candidiasis. In: GOLDMAN, L.; AUSIELLO, D. Cecil: Tratado de Medicina Interna. 22. ed. Rio de Janeiro: Elsevier. 2005; 359:2713-17.

KERN, M. E.; BLEVINS, K. S. **Medical mycology**: text and atlas. São Paulo: Premier, 1999.

LACAZ, C. S.; PORTO, E.; MARTINS, J. E. **Mycologia médica.** 8. ed. São Paulo: Sarvier, 1991. 693p.

LACAZ, C.S; PORTO, E.; MARTINS, J.E.C. *Medical mycology*: fungi, actinomycetes and algae of medical interest. 8 ed. São Paulo: Sarvier, 1991.

LANGONI, H; GUIMARÃES, F.F; WANDERLEY, G. G; et al. *Geotrichum candidum* and its importance in mastitis. **Veterinária e Zootecnia.** June, 2013 264-269.

PAULA, C. R.; RUIZ, L.S. Compênio de Micologia Médica: Técnicas para diagnóstico precoce das infecções por leveduras do genus cândida. 2 ed.

PERDONCINI, M. G. Microbiologia dos fungos, 3.ed. Rio de Janeiro, Moderna, 2006.

QUEIROZ. Annual report on research into alcoholic fermentation: 1991. Piracicaba, 1991. p.51-56.

RIBEIRO, E. L.; et al. Germ tubes in the genotyping of oral isolates of Candida albicans from children with Down Syndrome and their parents and/or carers. **Clipe Odonto- UNITAU.** v. 2, n. 1, p. 34-38, 2010.

SIDRIM J.J.C., ROCHA M.F.G. Micologia Médica à Luz de Autores Contemporâneos. Rio de Janeiro, Guanabara Koogan S. A., 2004.

VIEIRA, D.A. P. ; FERNANDES , N. C. A. Q. Microbiologia Geral, Caderno, Ciência e Tecnologia de Goiás- e-Tec Brasil.12 December 2007Goiás/IFG-Inhumas.

VIDOTTO, V. Manual de Micologia Medica. São Paulo: 2004. Tecmedd Publishing House.

William J D.; Berger, G Timothy G.; et al. (2006). *Andrews' Diseases of the Skin: **clinical Dermatology***, Saunders Elsevier. ISBN 0-7216-2921-0.

ZAITZ, C.; RUIZ, L. R. B.; SOUSA, V. M. **Atlas of medical mycology:** laboratory diagnosis. 2. ed.

CHAPTER 3

TECHNIQUES FOR IDENTIFYING FILAMENTOUS FUNGI

Henrique Barros Caminha
Kelly Maria Rêgo da Silva

Fungi exhibit micro and macroscopic morphological structures and are found in two groups: filamentous and yeasts. Filamentous fungi, which are also known as moulds, are multicellular, composed of hyphae, which associate and form the mycelium, resulting in filamentous colonies, which can be powdery, velvety or cottony and have a variety of pigmentation such as beige, white, burgundy, brown and others.

The identification process favours a broad knowledge and understanding of the morphology and function of the fungal population. Fungal identification techniques are immensely important because, based on the characteristics identified, it is possible to determine their role based on the data and information obtained.

Analysing a fungus in a biological sample has enormous diagnostic value, so the quantity of the sample must be taken into account along with its preservation. There are various methods for identifying and classifying filamentous fungi, such as direct microscopic examination of the sample, culture to isolate the fungus, microcultivation technique, among others.

Direct microscopic examination:

Direct examination or direct microscopic examination is one of the most widely used methods, it is sensitive and indispensable in the diagnosis of filamentous fungi. The material to be analysed may or may not contain fungal structures. There are several procedures for carrying out this direct examination, the difference being the solutions used, such as calcofluor white, India ink, KOH 10 - 40%, lactophenol blue. This examination consists of adding a small amount of the sample directly onto a slide and adding a drop of one of the solutions listed above, then taking it to the microscope and observing its morphological characteristics.

Culture to isolate the fungus:

The culture used to identify fungi is based on isolating the sample in a growth medium that favours the proliferation of the fungi to be identified and inhibits any other microorganisms that are not part of the study in question. The most commonly used medium is Sabouraud Destrose Agar (SDA), also known as Sabouraud Agar. As a rule, an antibiotic is used to prevent the growth of bacteria that can hinder the growth of fungi in the culture, chloramphenicol is the most suitable. After sowing the sample, the growth of the colony is observed in the oven at 37°C, and its growth can be classified as fast or slow. Isolation of filamentous fungi is carried out after direct examination, using plate or tube culture.

For a better identification, the colony must first be analysed macroscopically, either in a test tube or in a Petri dish (which is more recommended), and from these observe and identify its main characteristics in terms of texture, colour, growth time and surface, and then observed microscopically, where it will be analysed from a small amount taken from the colony in culture and added to a slide, then taken to the microscope to observe the presence of hyphae, which can be septate, continuous (cenocytic), dematiaceous or hyaline. Other important and satisfactory characteristics for identifying filamentous fungi are their morphology and spore formation.

Microcultivation technique:

To carry out this technique, some materials are used, such as a sterile slide, sterile Petri dish, ASD agar cube or potato agar, support, sterile coverslip, distilled water, formaldehyde, adhesive tape, tweezers, cotton wool and lactophenol blue dye.

The procedure consists of placing the agar cube on the sterilised slide that should be on the support, contained in a Petri dish, sowing the fungus from the recent repique on all sides of the agar cube and covering with the coverslip, adding two mL of sterile distilled water to the bottom of the dish to prevent the culture medium from drying out during the growth of the fungus, covering the dish and leaving it at room temperature for 8 to 12 days until hyphae appear.

The sporulation should then be inactivated by adding one mL of formaldehyde to cotton wool and covering the plate with adhesive tape for two days so that the formaldehyde helps to fix the microscopic structures.

Then remove the coverslip with tweezers and drip a drop of lactophenol blue colouring and mount on a slide. The hyphae and spores of the fungus should be adhered to the coverslip. Finally, discard the agar cube and drip another drop of lactophenol blue dye, covering with a coverslip to visualise the characteristics of the spores and hyphae also adhered to the slide.

To complement the identification of filamentous fungi, other more specific and sensitive tests are carried out to improve diagnosis, such as biochemical, biological and serological tests, as well as molecular biology diagnostics.

REFERENCES

ALMEIDA, MVA; Identification of filamentous fungi present in an urban solid waste bioreactor; Campina Grande-PB; 2015.

Mod VII; Detection and Identification of Fungi of Medical Importance.

MEZZARI, Adelina; FUENTEFRIA, Alexandre Meneghelo. Mycology in the clinical laboratory. Barueri, SP:Manole, 2012.

FAIA, Ana Margarida. Isolation and identification of filamentous fungi and yeasts in some points of a water distribution network. [dissertation]. Lisbon: Faculty of Sciences, Department of Plant Biology, University of Lisbon, 2011.

CHAPTER 4

ANTIFUNGIGRAMA

Kelly Maria Rêgo da Silva

The sensitivity test allows the *"in vitro"* determination of the sensitivity of certain fungi to various antibiotics. Based on this result plus the physical conditions, the clinician can choose the most appropriate antimicrobial therapy for the treatment of an infection.

KIRBY-BAUER method

This method is a CLSI standard sensitivity test. The plates and discs are left at room temperature before the antibiogram is carried out. Paper discs impregnated with a standardised concentration of the antifungal are used and placed on a Mueller-Hinton agar plate previously seeded with a standardised suspension of the microorganism. After incubation for 24 to 48 hours (depending on the microorganism), the inhibition halo is measured and interpreted as sensitive, intermediate or resistant, according to the criteria established in the CLSI document. After making the suspension, sow the seeds by inserting a sterile swab into the adjusted suspension within 15 minutes of adjusting the turbidity. Rotate the swab several times and squeeze it against the inside wall of the tube to remove excess inoculum. Then sow on the surface of the appropriate agar in 4 different directions. Immediately after placing the discs; use tweezers to place the discs on the surface of the plate, pressing lightly on each disc. Once the disc has been placed, do not remove it, as the drug diffuses immediately.

Reading the plates

After the incubation period, read the plates using a ruler or caliper. The inhibition halos for each antimicrobial tested should be interpreted in the sensitive, intermediate or resistant categories, according to the criteria established in the CLSI M100 tables (revised annually).

Sensitive (S): indicates that the infection due to a strain can be appropriately treated with the doses of the antimicrobial agent recommended for that type of infection and for the infecting species, unless contraindicated for some reason.

Intermediate (I): includes isolated strains with MICs similar to the levels normally reached in blood and tissues for which the magnitude of the response may be lower than for sensitive isolates.

Resistant (R): indicates that the isolated strains are not inhibited at the systemic concentrations usually achieved by the agent when administered at the normal dosage, and/or may have MICs that are within a value where specific microbial resistance mechanisms are likely, and if clinical efficacy has not demonstrated reliable results in treatment studies.

ANTIFUNGALS USED IN THE CLINIC

Antifungals used for systemic mycoses:

Amphotericin B

Amphotericin B is formulated for injection. This antifungal degrades the sterol portion of fungal membranes.

It has antifungal activity against mycoses caused by *Candida sp, Cryptococcus neoformans, Blastomyces dermatitis, Histoplasma capsulatum, Sporothrix schenckii, Coccidioides immitis, Paracoccidioides braziliensis, Aspergilus sp and Penicillium marneffei* and against agents that cause mucormycosis. Only two fungi are currently resistant to Amphotericin B, namely *Candida lusitaniae* and *Aspergilus terreus*.

Flucytosine

Flucytosine acts directly on the genetic material of the fungus.

It has antifungal activity against mycoses caused by *Candida sp, Cryptococcus neoformans* and some agents of chromoblastomycosis. Currently there are rare cases of resistance reported and they are only observed during the treatment of the disease.

Imidazoles and triazoles

These are clotrimazole, miconazole, ketoconazole, econazole, butoconazole, oxiconazole, sertaconazole, sulconazole, terconazole, itraconazole, fluconazole, voriconazole, posaconazole, isavuconazole and thioconazole.

Imidazoles and triazoles act directly on the cytoplasmic membrane.

It has antifungal activities against mycoses caused by *Candida albicans, Candida tropicalis, Candida parapsilosis, Candida glabrata, Cryptococcus neoformans, Blastomyces dermatitis, Histoplasma capsulatum, Sporothrix schenckii, Coccidioides sp, Paracoccidioides braziliensis, Aspergilus sp and Penicillium marneffei, Scedosporium apiospermum, Fusarium sp, Sporothrix schenckii, Candida* krusei *and* dermatophytes.

Reports of resistance to imidazoles and triazoles for *Cryptococcus neoformans, Aspergilus fumigatus, Candida albicans and Candida glabrata* due to prolonged therapy.

Echinocandins

The echinocandins are Caspofungin, Anidulafungin and Micafungin.

They are only sensitive to certain species of *Candida sp* and *Aspergilus sp.* They act directly on the cell wall.

Resistance to this drug depends on mutations in the microorganisms during treatment.

Griseofulvin and Terbinafine

Griseofulvin and terbinafine act on dermatophytes: *Microsporum sp, Epidermophyton sp and Trichophyton sp.* These drugs can be used topically or orally. They act on the mitotic spindle, preventing it from dividing. Their resistance depends on the invasion of new fungi at the treatment site.

Antifungals used for superficial mycoses

These are topical drugs used for superficial mycoses. They are ciclopirox, olamine, haloprogyn, tolnaftate, naphthyfine and butenafine.

REFERENCES

ANVISA. **Main Infectious Syndromes**. Module I.

FILHO, Lauro Santos. **Manual of Clinical Microbiology**. 3 ed. Editora Universitária, João Pessoa, 2003.

KONEMAN, Elmer *et al.* **Microbiological Diagnosis Text and Atlas**. 6 ed. Guanabara Koogan, Rio de Janeiro, 2008.

OPLUSTIL, C.A.; ZOCCOLI, C.M.; TOBOUTI, N.R.; Sinto, S.I. **Basic procedures in clinical microbiology**. 3ed. São Pulo. Sarvier, 2010.

OPLUSTIL, C.A.; ZOCCOLI, M. C.; SINTO, S.I.; MENDES, C.M.F. Clinical Microbiology: **156 questions and answers**. 1ed. São Paulo. Saivier, 2005.

GOODMAN & GILMAN. **As Bases Farmacológicas da Terapêutica -12ª** Ed. - Editora: ARTMED, Year of **Edition** 2015.

CHAPTER 5

PREVALENT MYCOSES IN PIAUÍ

Enio Victor Mendes de Alencar
Italo Barros
Mayra Nogueira Soares dos Santos Sara
Tamiris da Silva Costa Simaira Oliveira
Tamires Dayane de Oliveira Araujo Wendell
Braga

CANDIDIASIS

CLINICAL PRESENTATION AND SYMPTOMS

Candidiasis is characterised by the association of factors related to the host and its predominant microbiota, which varies from a commensal relationship to a parasitic one. Thus, the fungal genus responsible for the occlusion of candidiasis is *Candida,* which has numerous yeast species, found in the most varied body sites, such as: the oropharynx, mouth, skin, vagina, anus, urine and faeces.

Candida species are resident microorganisms of the female genital flora and are largely responsible for maintaining it. Thus, as pathophysiological changes occur in the host in question, the fungal population increases and the individual becomes colonised by this agent, which transforms from a normal resident, with a commensal (benefiting) relationship of asymptomatic action, to a pathological agent with a parasitic relationship of symptomatic action in the host.

According to its clinical aspects, candidiasis has different courses that are delimited according to the species in force, the virulence strain and the patient's immunological conditions. Thus, based on where the disease manifests itself, this pathology can be detected in four areas, formalising itself into four different types: cutaneous, mucosal, oral, cutaneous-mucosal and haematogenous.

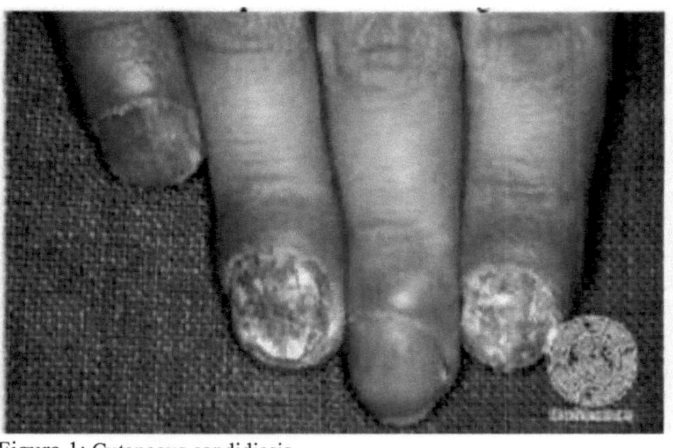

Figura 1: Cutaneous candidiasis
Sources: HTTP://www.dermis.net/dermisroot/pt/16108/image.htm. Brazilian Journal of
Surgery and Clinicai Research - BJSCR vol.8, n.2pp.75-82(Jun-Aug 2014).

The cutaneous form - is detected when the fungus in question affects the host's skin, the mucosal form - the mycological agent settles in the mucoid regions of the body. However, the oral form - the fungus manifests itself in the patient's oral cavity, and the cutaneomucosal form - the form that affects various regions of the body and is also related to the chronic form of the pathology, which is visceral candidiasis. Meanwhile, the haematogenous form - acts when the candida infection reaches the host's bloodstream in more severe cases.

In this way, we can emphasise that there are still other manifestations of the disease, in the acute and chronic phases of the infection, among which we can highlight the acute phase; vulvovaginal candidiasis and chronic phases; chronic vaginal candidiasis or recurrent vaginal candidiasis and visceral or disseminated candidiasis.

Vulvo-vaginal candidiasis, being an acute phase infection, is associated with imminent risk factors, such as: hormone replacement therapy, pregnancy without proper medical follow-up, use of high-dose oral contraceptives, decompensated *diabetes mellitus,* improper use of clothing that reduces air circulation in the genital area, improper hygiene, and other associated factors.

However, highlighting the chronic phase of the pathology, Recurrent Vaginal Candidiasis is categorically associated with infection through several sporadic episodes of the disease over a period of more than 12 months of vulvo-vaginal candidiasis (VVC).

However, a decrease in the natural and acquired immune response helps to reduce lymphocyte action for cellular defence, as is the case in patients with HIV (Human Immunodeficiency Virus) and hereditary immune pathologies, directly favouring immunoinfectious processes mediated by mycological agents.

However, there is a third category that is related to the spread of candidiasis and is determined by the presence of other pathologies or immunological conditions, such as leukaemia, immunosuppressive therapy, granulomatous disease and others. Where the fungus settles, it affects tissues and starts damaging processes.

Based on the utilitarian manifestations of the pathology, *Candida sp* has a preference for warm and humid places, and is therefore the main mycotic agent responsible for vaginitis and dermatitis. In addition, differentiating the pathogenicity of fungi can give them the necessary attributes to hatch a disease and cause an infectious process.

According to the distribution referring to the degree of differentiation of the pathogenicity of the fungi in question, we can highlight that the union with biological substrates, the production of germ tubes for the transformation of the filamentous form, antigenic variability, the destruction of the host's defence, and the production of toxinological agents, directly assist in the development of infectious processes in the body of the individual in question.

The yeast-like forms of the genus *Candida* can reach the female genitalia through the infection of fungi from the external environment, as well as through the over-colonisation of fungi from the body's normal microbiota. Thus, the symptoms of vulvovaginal candidiasis caused by *Candida sp* are clinically characterised by itching in the vulva, oedema and erythema in the female genital area and intense leucorrhoea.

CAUSATIVE FUNGUS AND AETIOLOGY

The main etiologic agent of candidiasis are yeast-like fungi of the genus *Candida,* among which we can highlight the species *C. albicans,* due to its ability to act as an opportunistic mycotic agent that resides on the mucous membranes of healthy individuals without causing signs of obvious disease under normal physiological clinical conditions. However, fungi from this species act in human colonisation after birth, and even in external infections.

Acting as a dimorphic fungus, *C. albicans* has a yeast or blastoconid form with asymptomatic infection and saprophytic formation from 37°C onwards. But it can also act as a filamentous or filamentous fungus from a temperature of 25°C, and can present itself as hyphae and pseudohyphae through processes with a high degree of infection.

However, the dimorphic fungi of this species in good conditions produce chlamydospores (circular spores with a rich, thick cell wall) and can adapt to different habitats, being characterised as a pleomorphic microorganism that can cause a high level of infections in the short and long term. However, in terms of yeast macromorphology, colonies can vary in colour from white to cream, with diffuse shiny and pasty aspects.

C. albicans is the main fungus causing vulvovaginal candidiasis due to its ability to adapt to high pH levels. Thus, this mycological agent has a PHR1 gene, which when active is associated with wall production and acts in conditions close to neutral pH. However, as the vaginal pH is not neutral but acidic, the expression of this gene is modified to produce a new gene like the first one, but which acts only at acidic pH to aid vaginal colonisation.

Over the years, there has been a steady increase in the appearance of *non-C.albicans* species, which are causing new infectious processes, acting asymptomatically in vulvovaginal candidiasis. However, the *non-albicans* species are highly pathogenic and have a higher level of resistance to antifungal drugs, which are azole derivatives.

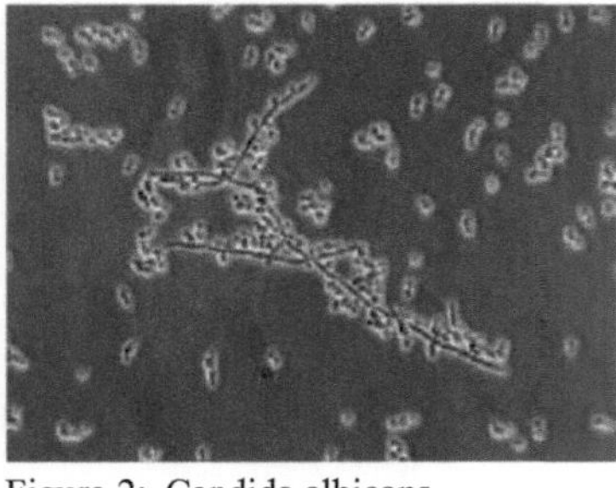

Figura 2: Candida albicans

Sources: HTTP://healthlob.com/2011/06/candidiasis-causes/ . Brazilian Jounrnal of Surgery and Clinicai Research - BJSCR vol.8, n.2pp.75-82(Jun-Aug 2014).

Species of the genus Candida:

Among the genera of *Candida,* the main causes of candidiasis and of interest in the clinic are: *C. albicans, C.tropicalis, C. guilhermondii, C.parapsilosis, C.glabrata, C.lusitaniae, C.krusei.* However, there are also cases of diseases that are related to some emerging fungal species, such as: *C.wickerbamii, C.kefyr, C.rugosa, C. lipolytica. C.norvegensis, C.inconspicua, C.famata, C.utilis, and C.dubliniensis.*

However, the major problem with *non-C.albicans* or *non-albicans* species is the dependence on high doses of antifungals to inhibit the pathogen and the recurrence of infections caused by these species. We can therefore classify these fungal species as inhabitants of the Kingdom Fungi, with Phylum Ascomyceta, Class Saccgaromycetes, Order Saccharomycetales, Family Saccharomycetaceae.

Candida albicans:

It is one of the main causative agents of candidiasis, thus developing superficial and invasive infections. They are assiduous residents of the microbiota of the human gastrointestinal tract. However, it is one of the most sensitive species to antifungal drugs, but there are indications that some strains of this species are resistant to classes of azole derivatives in prolonged use.

This species of fungus was one of the first to be genomically sequenced, thus enabling the discovery of alternative therapies and new experiments. As a zoopathogenic mycological agent, it has made a major contribution to advances in molecular biology techniques.

Candida dubliniensis:

It is one of the species most similar to *albicans,* but with less virulence. This type of fungus is a resident of the oral mucosa. But when it acts as a pathogen, it is associated with infections related to diabetic and immunosuppressed patients in the oral mucosa and oral cavity. However, *C.dubliniensis* still has hyphae, which gives it a lower level of invasion into the tissues, considerably reducing the risk of fungal colonisation.

As *C.dubliniensis and C.albicans have* similar properties, it is necessary to apply mycological tests to differentiate the two, such as: carbohydrate assimilation tests, macroscopic visualisation of the fungus using the Chromagar Candida culture medium (which has a greenish colour), chlamydospore synthesis, and gene sequencing.

Candida famata:

It is an opportunistic fungus that resides in the microbiota of the host's oral mucosa and can also be found in the exogenous environment in dairy products. However, according to the literature, it was believed that this fungus did not act as a cause of disease. However, it is now known that it can be related to retinopathies, fungemias and pathologies of the central nervous system. However, it is a mycotic agent that is poorly related to antifungals based on azole derivatives.

Candida ciferrii:

It is a yeast-like fungus that acts on soil and animals. However, in 1965 it was studied for the first time, with reports of onychomycosis and infections in HIV patients and cancer patients in the visceral form. *C.ciferrii* is a fungus that has similar characteristics to *C.curvata* and *C.hellenica,* as well as differing from *Cryptococcus* by urea hydrolysis tests.

Candida glabrata:

It can be considered a resident of the healthy human microbiota, but with the increase in the incidence of diseases such as AIDS and drugs, the incidence of this fungus has increased, making it the main promoter of hospital-acquired infections. However, based on the mortality associated with cases of candidiasis, *C.glabrata* is associated with high rates of candidiasis. However, it is important to emphasise that C.glabrata is not a dimorphic fungus and does not have pseudohyphae.

When it is associated with cases of animal infection, C.glabrata is less infectious. However, this type of fungus also pathogenically affects the oral mucosa, and when correlated with *C.albicans* in immunosuppressed patients, the infectious process in the oropharynx becomes greater and more difficult to treat. However, this type of infectious agent is less sensitive to fluconazole and amphotericin B.

Candida guilliermondii:

Although it is not very frequent in infections, it is related to onychomycoses, immunosuppressed patients and intensive care. *C. guilliermondii* is a haploid yeast, similar in structure to *C. famata,* which is divided into two other categories, C. *fermentati* and *C. carpophila.* However, the major problem with this fungus is that it is not very sensitive to many antifungal drugs.

Candida haemulonii:

As a mycotic agent, *C.haemulonii* is associated with cases of onychomycosis and vaginal candidiasis in immunocompromised patients. Visualisation of this fungus is not easy thanks to its phenotypic resemblance to *C.famata* and *C.guilliermondii.* However, studies have shown a lower sensitivity to fluconazole, intraconzaol and amphotericin B.

Due to mycological studies on *C.haemulonii,* a new species derived from this fungus has been discovered: *C.pseudo-haemulonii, which is* not sensitive to amphotericin B. However, *C.haemulonii* acts as an opportunistic fungus in certain types of pathology.

Candida inconspicua:

Acting as an opportunist, this fungus attacks patients with metabolic diseases such as Diabetes Mellitus and also those affected by viral, oral and vaginal infections. However, it has not been reported to cause cancer. Fluconazole has been shown to be an effective antifungal therapy against *C. inconspicua,* but amphotericin B is resistant to this class of fungus.

Candida kefyr

It is a frequent resident of the human microbiota, but when it acts as an opportunistic pathogen it is highly infectious and rare, causing opportunistic diseases. It also acts in different ways in dairy foods and industrial products.

Candida krusei

It is a highly resistant agent to fungicidal medication due to the incorporation of high doses of fluconazole and amphotericin B into antifungal therapy in immunocompromised patients. As a result, the only effective drugs to stop this type of infection are echinoclandins and vorinazole *in vitro.*

Candida lipolytica:

It is a fungal agent that acts on food products such as frozen meats, agricultural products, soil and vegetables. As an obligate opportunistic fungus, it affects the lungs, oesophagus, gastrointestinal tract and female genital system. Compared to *C.albicans,* it has less infectious potential and is sensitive to azole derivatives.

Candida lusitaniae:

Acting as an agent with great pathogenic potential, *C.lusitaniae is* resistant to amphotericin B, which is its distinguishing feature from other *Candida* species. In this way, it reproduces the switching mechanism that helps it to adapt to different locations in the human body and in drug resistance processes. Because of its similar appearance to *C. pulcherrima, it* is often confused with it due to its morphology. As a result, it is not recognised solely as C.lusitaniae, causing confusion between them.

Candida norvegensis:

It is a fungus that is difficult to treat due to its resistance to drugs based on azole derivatives and amphotericin B, and it is difficult to identify when compared to other fungal species such as *C.krusei* and *C.inconspicua.* Therefore, in order to differentiate these species, the process of hydrolysing esculin must be carried out, as well as diagnostic methods such as DNA studies.

Candida parapsolis:

It is a type of microorganism that comes from the body's normal microbiota. However, when it is pathogenic, it produces infections associated with the use of catheters, such as arthritis, endocarditis and fungal processes, and also acts on infections originating in the blood system. *C. parapsolis* is characterised by its subdivisions into other groups, such as *C. orthopsilosis and C. metapsislosis.* These fungi therefore have a reduced effect with the use of amphotericin B.

Candida rugosa:

Acting as a colonising fungus, it is associated with hospital contamination through catheters and high-risk infections in debilitated patients. Despite its resistance to polyene derivatives and fluconazole, it is a fungus with great biotechnological activity in food production and pharmacological development.

Candida tropicalis:

It is only associated with disseminated and opportunistic infections in debilitated patients, with high mortality rates, and with strains that are more virulent than those of *C.albicans*. However, it is highly sensitive to antifungals of the polyene class and azole derivatives.

Candida utilis:

With great importance in industry through the production of organic metabolites based on non-alcoholic fermentation, it is a rare pathogen with little virulence, acting on weakened hosts.

ASSOCIATED FACTORS

Fungi of the genus *Candida* are functionally harmful organisms to the human host when they are not in their pathologically colonising form. However, there are eminent risks for this type of microorganism to move from its natural condition to its pathogenic condition. These risk factors include the use of medication, women's physiological cycles, metabolic and immunological diseases, inappropriate basic hygiene habits and other factors.

Respectively, taking into account the first condition for *Candida* infections, the indiscriminate use of medication is promising for the increase in fungal colonisation, such as the practice of antibiotic therapy which eliminates a large percentage of beneficial bacteria from the body and alters the nutritional competition between fungi and bacteria in the individual's body, thus increasing fungal colonisation in the host.

In addition to the practice of antibiotic therapy, there is also the major issue of the use of oral contraceptives and hormone replacement therapy, which are drug-based mechanisms that increase the glycogen peak, favouring nutrients for mycological agents so that they can increase colonisation of the genital mucosa, thus facilitating the development of infections.

However, injectable contraceptives provide greater protection against mycotic agents than oral contraceptives, as they have low levels of oestradiol during the lactation phases of a woman's cycle. However, there is no evidence that candidiasis is associated with the IUD - Intrauterine Device.

The use of corticoid-based medicines is also widely associated with fungal infections. Since these drugs compromise the immune system of the patient who uses them over the long term, opportunistic fungi such as *Candida sp.* act and find the host's defences compromised, so the fungus settles in, destroys tissues and causes damage to the individual s body.

According to the second condition, a woman's physiological state also favours the appearance of mycological agents, which act on her menstrual cycle, especially in regular cycles due to the high levels of oestradiol during this period. However, pregnancy also increases the risk of candidiasis, because as a woman's body undergoes hormonal and physiological changes, infections of the vaginal mucosa are more prevalent.

However, with regard to the prevalence of candidiasis, the risk of HIV (Acquired Immune Deficiency Syndrome) in pregnancy, as well as other sexually transmitted diseases (STDs), favours the appearance of these fungi.

These cases are linked to the side effects of pregnancy, such as ruptured membranes during labour.

On the other hand, the third condition associates the prevalence of candidiasis with metabolic diseases, such as decompensated *diabetes mellitus,* which causes an increase in glycogen peaks that help the pathogenic condition of *Candida sp.* However, when the disease is controlled and healthy habits are promoted, the colonisation of the fungus decreases, as do the infectious processes resulting from it.

However, there are still other diseases that can lead to a decreased immune response in the individual's body, such as the alterations associated with HIV patients, cancer patients and those with hereditary immune defects. However, the common factor for the involvement of this fungus in these pathologies is the decrease in the immune response due to the disease and the invasion of opportunistic microorganisms.

To highlight the fourth condition, according to habits we can highlight the use of tight clothes or those made with synthetic fibres, which compromise air circulation in the genitals, cause an increase in humidity and favour the appearance of this fungus. Maintaining inappropriate hygiene habits such as anal-vaginal cleansing and even some sexual practices also help in the process of fungal colonisation.

Among the other factors that can trigger the appearance of candidiasis are those related to the fungus being introduced into the bloodstream, through the administration of venous and intravenous access, drug abuse with the use of syringes, dialysis processes, the introduction of catheters and even surgical processes without proper sterilisation of materials. This is because *Candida sp* fungi can live in various places and maintain themselves in various temperature conditions.

DIAGNOSIS

For the diagnosis of candidiasis to be made effectively, the patient must first be guided through the clinical process, which extends from a medical consultation with the appropriate anamnesis to laboratory diagnosis. However, it is important to emphasise that each diagnostic methodology is modified on the basis of the types of candidiasis and the symptoms that are detected in the patients in question.

Therefore, in daily clinical practice, it is important to differentiate what is observed in various aspects, such as: the findings of *Candida sp* in routine gynaecological examinations, from patients who present symptoms similar to the fungus in question but who have not undergone any laboratory procedure to confirm it.

Therefore, taking into account the first aspect in differentiating the clinical diagnosis, we must demystify the issue that every laboratory finding in routine tests such as the Pap smear that determines the presence of *Candida sp* is already the outbreak of candidiasis. Of course not, because the Pap smear is not a confirmatory test for the disease in question and as it is invasive, it can reach the vaginal mucosa where the normal microbiota of the vagina reside.

Based on the second aspect, it is important to determine that there are several diseases that affect the human body that have similar symptoms. But it is important to choose a confirmatory diagnostic method to study them. However, if there are no clinical symptoms of the disease and other tests still show these pathogens, a more detailed analysis of the finding is necessary.

Mycological tests for assimilation, sugar fermentation, morphological visualisation of colonies, cultivation of samples using chromogenic media and species differentiation techniques have traditionally been used to diagnose *Candida sp.*

In order to help and reduce the limitations of phenotyping and immunophenotyping, effective molecular biology methods have emerged that have been adapted to detect *Candida sp* based on *Goebel'*s standardisation of a species-specific PCR. The use of this species-specific method based on the amplification of transcribed sites of an rRNA gene of the fungus in question has considerably improved the diagnosis of *Candida sp.*

To achieve this, 7 species-specific primers were used for amplification, plus 1 universal primer that binds to the 28S rRNA gene site. According to the sequence formed, these primers help to determine various species of the fungus, such as: *C. lusitanie, C. parapsilosis, C. tropicalis, C.albicans, C.glablata, C.krusei,* and *C.guilliermondii.*

The variation in the behaviour of some *Candida* species is also evaluated in some diagnostic procedures, as it is necessary to use some practical and rapid methodologies to detect them, such as the use of selective culture media for fungal growth and even the use of enzymatic panels and automated equipment.

PREVENTION

The basic methods of prevention against *Candida sp are* to take into account the risk factors and even avoid some of them. For example, having sex with a condom and having a gynaecological preventive examination every 6 months in women, because even though this examination is not confirmatory for *Candida* species, it is effective in preventing an increase in fungal colonisation.

It's also important to wear less tight-fitting clothes to give the genitals more air circulation and reduce humidity in these areas. Basic hygiene habits and a healthy diet with carbohydrates, vitamins and proteins are also essential for reducing infections by these mycological agents, in order to keep the immune system in full vigour and avoid diseases that cause opportunistic agents to enter the body.

TREATMENT

Treatment of candidiasis is based on mycological drugs, such as imidazole and triazole drugs: miconazole, fluconazole, clotrimazole, itraconazole and others based on ketoconazole. Polyene derivatives such as amphotericin B, nystatin and others are also very important.

However, there are various *non-albicans* species that are not sensitive to antifungal treatments, making it difficult to treat them. However, as a way of minimising the action of these agents, the *National Committee for Clinical Laboratory Standards* (NCCLS) has developed the M27-A methodology for standardising susceptibility tests to aid in diagnosis and treatment. However, there are also other in vitro techniques that are similar to the M27-A and are simple to use, such as the *Etest.*

ASPERGILOSIS

INTRODUCTION

Aspergillus spp. is a fungus belonging to the Aspergillaceae family that has a universal distribution. It is generally found in soil, air, water and even in any decomposing matter, facilitating the dispersal of conidia, which is its infective form. Its main source of infection is the airborne route, affecting mainly immunosuppressed patients, children and the elderly, as the clinical manifestation of this disease is directly related to the host's clinical response. The species that most affect humans are: Aspergillusflavus(5-10%), Aspergillusnidulans,Aspergillusniger(2- 3%),Aspergillusterreuse the main one responsible for infections in humans is Aspergillusfumigatus(85%) - (Image 1).

Image 1-Image of Aspergillus fumigatus in electron microscopy, Taken from the master's thesis SPERGILLUS AND ASPERGILOSIS - CHALLENGES IN COMBATING THE DISEASE. site:http://www.niaid.nih.gov/dir/labs/lci/aspergillus.gif (11-10-13).

Aspergillus fumigatus:

It was first described in 1863 by the physician Georg W. Fresenuis, the name derives from the Latin "fumigave" referring to the smoky blue-grey mycelium. Its spores are produced in conidiophores that vary from 1.0 to 4.0 micrometres in diameter. Regarding its macroscopic appearance, it generally has velvety colonies with a bluish-grey or greenish-grey colour, with pigment production rarely occurring in some samples.

It is a thermotolerant fungus and grows at temperatures ranging from 15°C to 53°C. It has the ability to cause various types of infections and can affect the eyes, cardiovascular system, central nervous system and lungs, causing invasive infection and pneumonia in hosts with a depressed immune system (Image 2).

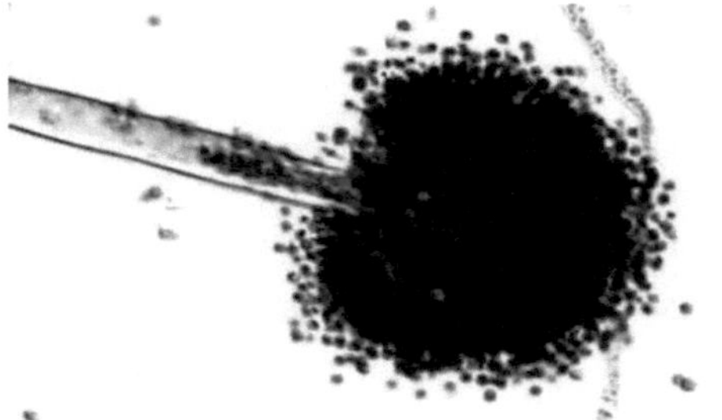

Image 2 - Macroscopic appearance of Aspergillus fumigatus colonies in Sabouraud medium Dextrose Agar, incubated for 5 days at 25°C. Taken from the atlas of medical mycology.

Aspergillus niger:

It was first described in 1867 by Van Tiegher and has the ability to develop colonies in less than seven days. At first it looks like a yellowish or white suede, but it is quickly covered by a layer of black conidial heads. The finding of calcium oxalate crystals in samples is suggestive of Aspergillusniger infection (Image 3).

Image3 -Microscopic view: Aspergillus niger aspergillus head. Taken from the website: http://www.pfdb.net/html/species/s12.htm (11-10-13).

Aspergillus terreus:

Identified by Thom in 1918, it has the ability to develop colonies with a suede, cinnamon-coloured appearance, has an incubation period of 7 to 14 days, is the main cause of allergic or invasive broncho-pulmonary infections, cutaneous and disseminated mycoses, is of great concern due to its resistance to Amphotericin B, Aspergillusterreus infections result in a higher mortality rate compared to other infections caused by other *Aspergillus spp.*

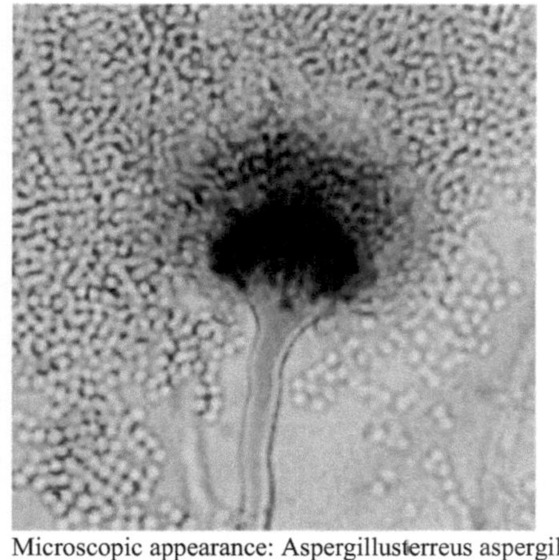

Microscopic appearance: Aspergillusterreus aspergillus head. Taken from site:http://www.mycology.adelaide.edu.au/Fungal_Descriptions/Hyphomycetes_(hyalne)/Aspergillus/terreus.html (11-10-13).

MORPHOLOGY

Fungi of the genus Aspergillus are characterised by the presence of numerous septate hyphae forming a filamentous stalk, with dichotomous branching forming an angle of 45°. Macroscopically, the colonies are white in their initial phase, but this can vary according to the species, evolving to yellow, green, black or brown with a cottony texture, with roughness on the wall due to the large production of spores.

PATHOGENESIS AND PATHOLOGY

Infection occurs due to the inhalation of Aspergillus spores dispersed in the air, the incubation period of which is unknown. The conidia are inhaled, enter the respiratory system and germinate, giving rise to hyphae in the lungs that will later invade other tissues and organs.

Some factors can favour infection by Aspergillus, the main pathogenicity factors being: Ability to proliferate at 37°C - *body temperature* -, small size of conidia, favouring dispersal through the air, ability to invade and adhere to endothelial tissue and exacerbated production of toxic extracellular products such as: elastase, fumigatoxin, restrictocin among others.

PREVENTION

It's practically impossible to avoid the fungus completely, but there are some precautions you can take to avoid possible contact with the fungus Aspergillusssp. If you have a respiratory problem, asthma or a low immune system, you should avoid very humid places, forests, compost heaps, and if you really need to go into these places, it's recommended that you wear a mask.

DIAGNOSIS

Aspergillosis is difficult to diagnose because the signs and symptoms are somewhat non-specific, and often the symptoms are similar to those of other fungal infections. For good sensitivity and specificity in diagnosis, it is necessary to use several techniques together: laboratory tests, microscopy tests through direct analysis of tissue sections and clinical samples, culture of the fungus, diagnostic imaging tests such as teleradiography and serological tests such as immunological, molecular and biochemical markers, which are the fastest, most specific and sensitive in detecting a possible infection.

TREATMENT

There are three classes of drugs most commonly used in the treatment of fungal infections: polyenes, azoles and echinocandins, which are ideal because these drugs have different mechanisms of action on different targets.

Polyene derivatives

Amphotericin B is a drug from the polyene class and is currently the antifungal drug with the broadest spectrum of action. Its action depends on binding to the ergosterol present in the fungal membrane. Immediately after binding, ion channels are formed in the membrane, causing a change in the cell membrane and the loss of constituents and subsequently cell death.

Azole derivatives

There are two main groups, the imidazoles and the triazoles, both of which have the same mechanism of action. Of the azole class, only voriconazole, irtaconazole and posaconazole have antifungal activity against *Aspergillusssp,* all of which have good bioavailability when administered orally.

Its mechanism of action consists of inhibiting ergosterol biosynthesis by inhibiting the enzyme 14-a-desmethylase of lanosterol, an enzyme that is directly linked to the conversion of lanosterol into ergosterol.

Echinocandins

They are basically used against infections caused by Candida and *Aspergillus. They do* not have good bioavailability when administered orally, which is why they are used intravenously.

Its mechanism of action basically consists of inhibiting the synthesis of glycans in the fungal wall by inhibiting the enzyme 1,3-Pglycansmtase.

CRYPTOCOCOSE

Cryptococcosis is a mycotic disease caused by the basidiomycete genus Cryptococcus. It is an opportunistic mycosis that has an affinity for the central nervous system and the integumentary system. It is also known as torulosis, European blastomycosis and Busse-Buschke disease. Its etiological agent is *Cryptococcus neofarmans,* the main human pathogen among the 37 species present in the genus, but currently there are two species: Cryptococcus *neofarmans* (serotypes A and D) and *Cryptococcus gatti (serotypes* B and C).

It is a yeast encapsulated by a mucopolysaccharide, globose or oval with a diameter of 3 to 8 pm. It has white to cream-coloured colonies with a mucoid texture and can be cultivated on media such as Sabouraud Agar and Malt and Yeast Extract Agar. Its infection occurs naturally in humans and animals through the inhalation of dehydrated yeasts in the environment in the droppings of pigeons or other birds. These droppings are rich in urea and creatinine, which favour the growth of this type of yeast.

These airborne propagules are inhaled through the respiratory tract until they reach the lungs, characterising the pulmonary form of the disease, which can spread haematogenously to other organs and tissues. Its pathogenic action depends mainly on the capsule, which contains glucuronoxylomannan that is present in all serotypes. Its antigenic and adhesion capacity is capable of inhibiting phagocytosis, consuming the complement system and neutralising opsonins and protective antibodies, thereby inhibiting leukocyte migration.

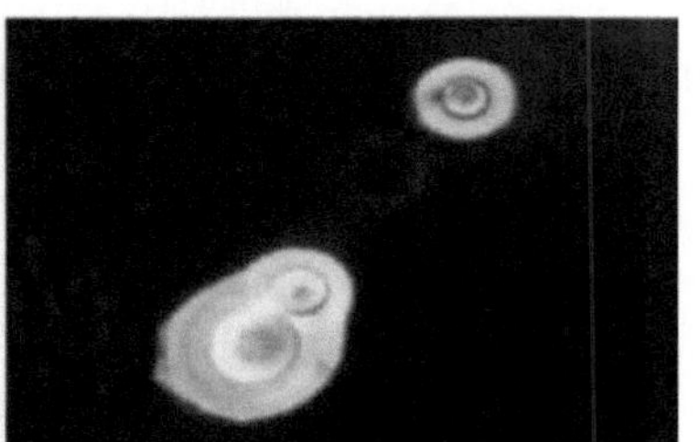

Figura 1:   Microscopic image of *Cryptococcus* stained with India ink. Source: Consensus on cryptococcosis, 2008.

Clinical signs can vary according to the organ affected, which can generate local manifestations or spread to cause respiratory failure. These manifestations are usually related to the central nervous system, skin, lymph nodes, bones/joints, eyes, liver, heart, kidneys and others.

One of the most common manifestations is subacute and chronic meningoencephalitis, often in immunocompromised patients and sometimes in immunocompetent patients after pulmonary involvement. The pulmonary lesion can be asymptomatic in 1/3 of cases, or symptomatic when it presents clinical signs such as sputum production, fever, chest pain, cough, night sweats, dyspnoea and obstruction of the superior vena cava.

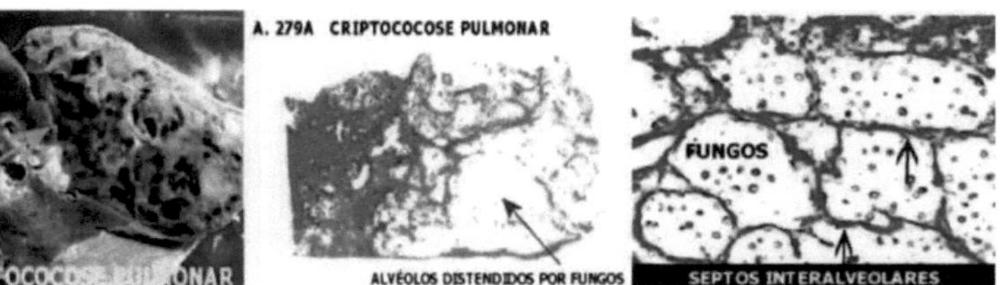

Figura 2:   Pulmonary cryptococcosis - Lung involvement
Source: www.scielo.br (Consensus on cryptococcosis, 2008)

Skin lesions are seen in 10 to 15% of cases, occurring by haematogenous dissemination (secondary cryptococcosis) and the rarer form by localised inoculation into the skin (primary cryptococcosis). Skin manifestations can include papules, infiltrated plaques, pustules, nodules, oedema, abscesses or ulcers.

DIAGNOSIS

The diagnosis of cryptococcosis is usually based on the patient's clinical picture and history. Laboratory diagnostics can make it easier to recognise this disease, using materials such as cerebrospinal fluid (CSF), urine, tissue fragments, aspirates from skin lesions, sputum and other samples.

Direct examination, cultures, molecular and serological tests can be carried out. Direct examination tests for the pathogen using secretions, sputum, cerebrospinal fluid (CSF) and ganglia, stained with 10% KOH and India ink or nigrosin, which reveals globose yeasts with single buds and polysaccharide capsules. For culture, the material is inoculated onto Sabourand glucose agar with chloramphenicol, BHI agar and differential media such as Niger seed agar, sunflower seed agar, dopamine agar or carrot potato agar.

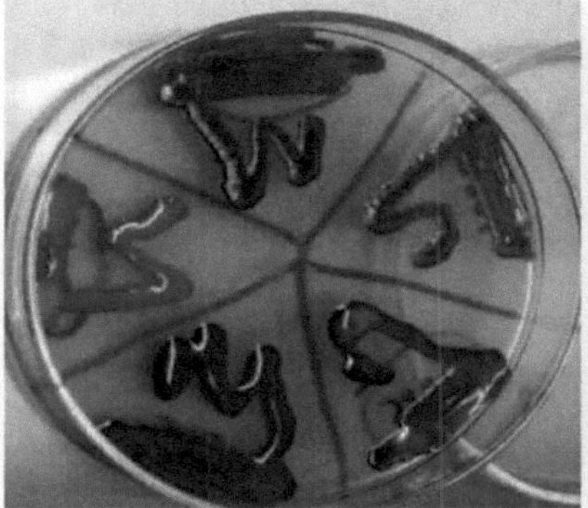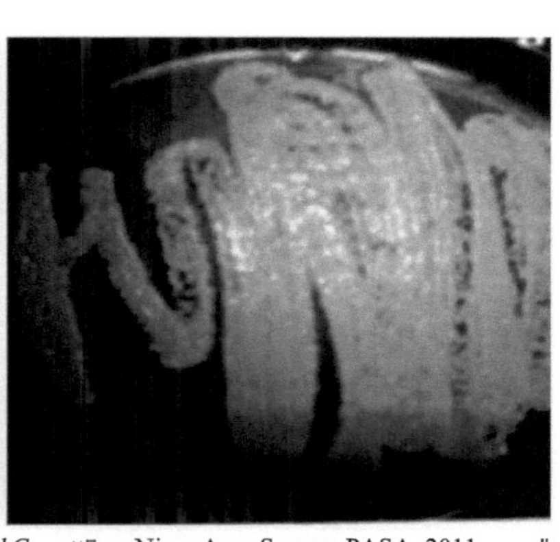

Figure 3: Cultivation of *Cryptococcus neoformans and C. gattii* on Niger Agar Source: PASA, 2011 " "
Figure 4: Cultivation of C. neoformans in DSA Source: FARIA, 2010

These media produce an oxidation reaction due to the micro-organism producing a phenoloxidase enzyme that acts on the substrates or polyphenoic compounds, thereby producing melanin, which results in a brown or black colouration of the colonies. Its evolution shows a positive and capsulated urease and inositol reaction, making it suspicious for *Cryptococcus neoformans*.

In molecular tests, identification methods are used to diagnose the serotype and strains. One of the most sensitive and specific techniques is PCR (Polymerase Chain Reaction), which analyses cerebrospinal fluid samples with suspected neurocryptococcosis. In serological tests, diagnosis is quick and safe because it has high sensitivity and specificity when testing for the circulating polysaccharide antigen in serum or cerebrospinal fluid, using latex agglutination, where titres of 1:4 are already suggestive of cryptococcal infection and titres of 8 or more indicate active disease.

PREVENTION

Cryptococcosis is a cosmopolitan disease that affects both animals (cats, horses, cows) and humans. There are currently several ways to reduce or even prevent human contact in contaminated areas. Health education is a great way to do this, as it aims to educate society about the forms of transmission and means of prevention.

As the fungus is found in soil, fruit, cow's milk and droppings, the main preventative measures are local humidification, preventing the propagules from dispersing into the air and controlling the increase in the pigeon population. It is necessary to wash areas contaminated by these faeces with chlorine, wearing gloves and masks to avoid direct contact with the dried droppings.

RISK FACTORS

Its incidence has increased significantly with the advent of transplants and AIDS.

Cryptococcosis associated with AIDS has a high lethality rate, which is being reduced by antiretroviral therapies. There is also a high incidence of this infection in patients with autoimmune diseases and transplant patients.

Cryptococcosis occurs most frequently in males and adults.

TREATMENT

Existing therapies vary according to the severity of the infection and the immune status of the host. In immunocompetent and immunocompromised patients, amphotericin B associated with 5-flucytosine is used in cases of disease dissemination, while fluconazole and itraconazole are used in skin lesions.

In the case of seropositive patients with cryptococcosis, amphotericin with or without 5-flucytosine is recommended, consolidating therapy with fluconazole for 10 to 12 weeks. It is necessary to have three (3) negative CSF cultures at one-month intervals to be considered cured of the disease. It is possible to present resistance to certain drugs, in the case of fluconazole used in therapies with seropositive patients with *Crytococcus neoformans* meningitis, there are cases of clinical non-improvement.

Combined fluconazole (400 to 800 mg/day + 5-FC (150 mg/kg/day) has shown effects in the treatment of cryptococcal meningitis in AIDS patients.

Data indicates that amphotericin B has reduced the mortality rate of cryptococcosis by 30 per cent. Therapy can be discontinued in cases where patients no longer show clinical manifestations, negative cultures, decreased titre of cryptococcal antigen in CSF and blood.

HISTOPLASMOSIS

Histoplasmosis is a fungal disease that affects the lungs in particular, and its pathogenicity is related to the host's immunological status and the acquired fungal load. It is regularly studied as an asymptomatic or self-limiting infection, but can evolve into the following proportions: acute pulmonary histoplasmosis, chronic pulmonary histoplasmosis and disseminated histoplasmosis.

The disease was first reported by Samuel Darling in Panama, who between 1905 and 1906 autopsied three widespread cases of the disease, two of which were from the island of Martinique, where this mycosis is now recognised as endemic.

CAUSING FUNGUS

The causative agent is Histoplasmacapsulatum, a dimorphic fungus with two types: *H.*

capsulatum var.capsulatum (responsible for classical histoplasmosis) and *H. capsulatum var. duboisii* (responsible for African histoplasmosis).

H. capsulatum is often detected in soils in its saprophytic or filamentous form. When incubated at 37°C it takes on the parasitic form, which consists of rounded or oval yeast-like elements.

There is a specific association between histoplasmosis and the existence of birds. Birds don't harbour the fungus because their high body temperature doesn't allow it to grow, but soil that has benefited from stools allows these organisms to grow. Bats, being mammals, can receive histoplasma in their digestive tracts and disperse it from their habitat, which includes caves, basements and holes in trees.

Once the fungus has settled in the soil, it continues to spread extensively, even if the area is devalued by birds. In contaminated soil, most of the infectious particles are found in the most obvious layer and are spread mainly by the wind.

ETIOLOGY

Two forms of *Histoplasma capsulatum* are related today: varcapsulatum and var duboisii. The two are indistinguishable in their mycelial form, but differ in their yeast-like form; in var duboisii, the cells are larger and have more condensed walls than in var capsulatum. On Sabouraud agar, at room temperature, the mycelium grows in the form of a white aerial colony, which englobes microconidia, as well as tuberculated macronidia, the latter covered by spiculated projections.

The mycelium is hyaline, branched and septate, 2 to 4 metres in diameter, and the microconidia are pyriform and smooth-walled, measuring 2 to 5 metres in diameter. The sexual phase is called Ajellomyces capsulatus and its isolation in the clinical laboratory is not common enough. The fungus can grow well on various media other than Sabourand agar, such as potato-glucose agar, Borelli's lactrimel or earth extract agar.

At 37°C, the yeasts are small (3 to 5 metres in diameter), oval and regularly have single twinning. These yeast-like elements can be observed inside macrophages or giant cells and more infrequently inside polymorphonuclear neutrophils.

In the Giemsa (or Wright) stain, fungi show a polar, blue, half-moon-shaped colour mass. PAS stains them red and Grocott's methenamine silver stains them black or dark brown.

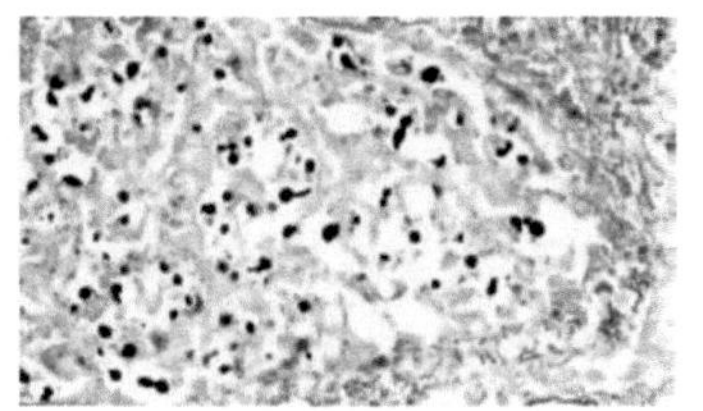

Image: 1 - Yeast forms of *Histoplasmacapsulatum stained* with silver (Gomori- Grocott, 400 x). Source: Ferreira M S, Borges A S. Histoplasmosis. Revista da Sociedade Brasileira de Medicina Tropical 42(2):192-198, Mar-Apr, 2009.

RISK FACTORS

The risk of infection is deeply related to the type of activity created by the subject, the time and regularity of exposure to the environment potentially contaminated with infectious particles, the fungal load, the virulence of H. capsulatum, as well as the individual's immunological status and genetic profile.

Systemic infection, known as disseminated histoplasmosis, is more common in patients who have some form of cellular immunity deficiency, such as human immunodeficiency virus-positive patients, patients on chronic steroid and other immunosuppressant drugs and people with haematological malignancies; or those who have undergone organ transplants, individuals of extreme age (under 1 year old and over 60 years old) and people with immunity deficits such as patients on leukaemic corticosteroid treatment and transplant patients.

Most infections are asymptomatic because they affect immunocompetent individuals with little exposure to the fungus.

SYMPTOMATOLOGY

Histoplasma infection occasionally results in varied and unusual clinical syndromes. Systemic symptoms are non-specific and include fatigue, fever, night sweats, anorexia and weight loss. Pulmonary symptoms include a productive cough and dyspnoea, similar to that seen in chronic obstructive pulmonary disease.

Histoplasmosis, in its acute forms, is a disease that regresses spontaneously. The asymptomatic or mildly symptomatic form is the most common, which often goes unnoticed because it is mistaken for the flu.

The so-called acute or epidemic form of pulmonary histoplasmosis can appear to clinicians as isolated cases, which are difficult to diagnose, or in the form of microepidemics, which are easier to diagnose and have a benign course, the symptoms of which depend on whether or not the patient is exposed to the infecting propagules. Fever, persistent unproductive cough, headache, asthenia, retrosternal pain and intense prostration are common. Skin paleness is a striking sign.

Enlarged superficial lymph nodes and hepatosplenomegaly are characteristic findings of the acute diffuse pulmonary form. Physical pulmonary signs are unexpressive.

PREVENTION

Although it is considered approximately impossible to manage the access of individuals to areas with a potential for H. capsulatum infection, it is considered necessary to understand the risks involved.

As preventative measures, it is suggested that health agencies, such as the Epidemiological Surveillance Groups, inform both the population and travel agencies that organise ecotourism or rural leisure activities about the need to wear a mask when entering caves, as well as that the collection and transport of soil, stones, plants and animals are not recommended, as they are potential sources of infection.

LABORATORY DIAGNOSIS

For an accurate diagnosis, the importance of anamnesis should be emphasised. This should be based on the patient's state of health and whether they have a history of living in an endemic country.

The recognition of a dimorphic fungus in any clinical material has a diagnostic connotation and it is necessary to obtain an adequate sample. Therefore, for pulmonary lesions with clinical suspicion of mycosis, specimens obtained by non-invasive methods such as sputum are first evaluated, followed by fibrobronchoscopy, bronchoalveolar lavage, transbronchial biopsy, transcutaneous biopsy and lung biopsy.

MYCOLOGICAL DIAGNOSIS

The isolation of H. capsulatum in culture is the gold standard for mycological diagnosis. It is important to emphasise that cultures on selective media should be part of the laboratory routine for potentially contaminated samples. The fungus is dimorphic, i.e. it manifests a mycelial form at 25°C and a yeast-like form when grown on rich media at 37°C11.

At 22-28°C, white, cottony, slowly growing cultures grow, with aerial mycelium that tends to darken over time. Microscopically, you can see delicate, septate hyphae, smooth microconidia or smooth, cornered chlamydoconidia, as well as a large number of tuberculate (or mammillate) macroconidia known as stalagmospores.

In rich media such as BHI, with or without blood incubated at 37°C, Histoplasmacapsulatumvar.capsulatum grows in yeast form, forming creamy, moist, shiny and smooth colonies.

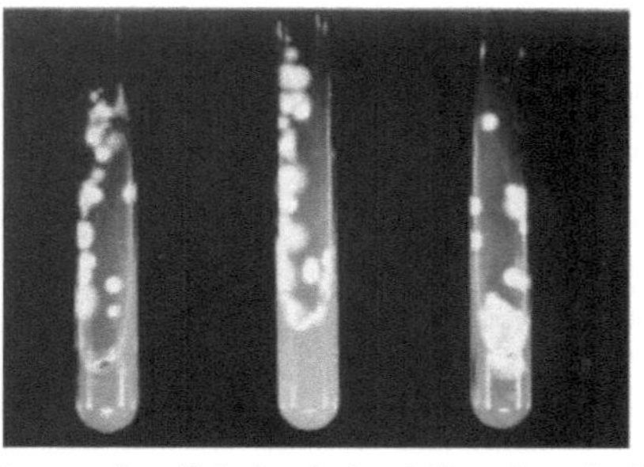

Image: 2 - Colonies isolated from bronchoalveolar lavage in Mycosel®, in the filamentous phase of Histoplasmacapsulatum var capsulatum, cream colour, at 25°C.Source:Unis G, Silva V B, Severo L C. Disseminated histoplasmosis and AIDS. Importance of the culture medium for the clinical-bronchoscopic specimen. Revista da Sociedade Brasileira de Medicina Tropical 37(3)234-237, mai-jun, 2004.

HISTOPATHOLOGICAL EXAMINATION

Histopathological studies of various tissues (lungs, ganglia, liver, bone marrow) show the presence of granulomas, with or without caseous necrosis, in immunologically deficient organisms, while in immunosuppressed organisms the presence of loose granulomas, lymphohistiocytic aggregates or just diffuse mononuclear infiltrates is common. The fungus in yeast form is seen inside the macrophages and also outside them.

SEROLOGICAL TESTS

Serological tests are very useful for diagnosing histoplasmosis, as they are positive in 80% of disseminated forms, 90% of acute pulmonary forms and 100% of chronic pulmonary forms.

There are two tests for evaluating the antigenic response to H. capsulatum: the ID on double agar gel and the FC test. The ID is a simple test and more widely available in medical practice than the FC. It identifies the M and H precipitation bands of the fungus. The M band can be detected in 75% of patients with acute histoplasmosis and almost all of those with chronic histoplasmosis; the H band is present in only 20% of them.

The H band indicates disease activity and the M band indicates recent contact with the fungus. Precipitins appear 4-8 weeks after exposure and after the appearance of complement-fixing antibodies.

Around 95% of patients with acute and chronic histoplasmosis are positive for the CF test; however, 25% of them are weakly positive, with titres of 1:8 or 1:16. Even low titres in a third of cases represent active disease. Although a four-fold increase in titres strengthens the significance of serological tests, they occur in 37% of patients with histoplasmosis.

RADIOGRAPHIC EXAMINATIONS

On plain radiography, according to the current consensus, in the chronic pulmonary form, there may be pleural reaction, fibrosis and contraction of the affected lobe or segment with a change in the diameter of the chest and deviation of the mediastinum. Involvement of lymph nodes adjacent to the bronchi by histoplasmosis can cause collapse of the middle lobe or other lung segments.

Computed tomography (CT) scans show the presence of nodules located in the intra- and interlobular region of the lung, which may have a perivascular distribution.

TREATMENT

Histoplasmosis is generally a benign, self-limiting infection. Cases of spontaneous regression do not require specific treatment, with rest and clinical observation being the most effective measures.

Ketoconazole is the treatment used for mild or moderate infection. In disseminated disease, systemic treatment with amphotericin B is almost always therapeutic, although some patients may require prolonged treatment and monitoring due to the occurrence of relapses.

AIDS patients relapse despite therapy, which would be curative in other individuals. AIDS patients therefore need maintenance therapy with oral ketoconazole or amphotericin B on a weekly basis.

COCCIDIOIDOMYCOSIS

CLINICAL ASPECTS AND CAUSATIVE FUNGUS

It is a systemic infection caused by the dimorphic fungus *Coccidioides immites, which* affects humans and animals. It is a fungal disease acquired by inhaling the agent *Coccidioides sp. in the* form of arthroconidia. It is not a serious infection, i.e. it resolves spontaneously. However, a small number of individuals have had progressively fatal cases, i.e. it has reached other organs besides the lungs through haematogenous dissemination.

Through molecular phylogeny studies, the existence of another species of etiological agent was discovered, and it was named Coccidioidesposadasii, in honour of its discoverer Alejandro Posadas. *Coccidioides immites* is compatible with fungi isolated in California, especially in the San Joaquin Valley (USA), and the *Coccidioides sp* species can be found in all the other endemic regions of the American continent. The agent of Coccidioidomycosis in Brazil is *C. posadasii.*

It is a mycosis found in arid and semi-arid regions of the American continent between parallels 40°N and 40°S, mainly in the south-west of the United States, northern Mexico and the semi-arid region of north-eastern Brazil. *Coccidioides sp.* is related to semi-arid environments, with high temperatures in the long dry season or scarce rainfall concentrated over a short period of time. It is therefore a disease with a limited geographical distribution and its transmission is limited to a few months of the year.

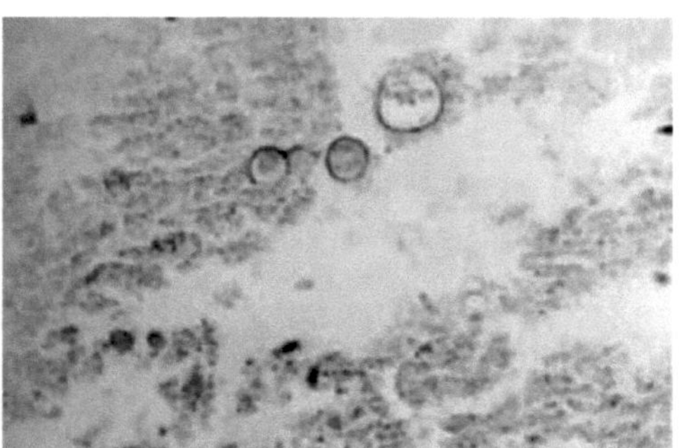

Figure 1: Lung necrosis with C.immitis spherules. HE 200x

Source: Revista da Sociedade Brasileira de Medicina Tropical 31:559-562, Nov-Dec, 1998.

ETIOLOGY

The occurrence of systemic mycoses in the semi-arid Northeast requires studies that can consistently demonstrate the epidemiological indicators of incidence, prevalence, morbidity and mortality. 10,000 new cases of infection are expected every year in the United States, 35,000 of them in California alone.

The mycosis has already been identified in four Brazilian states: Piauí, Maranhão and Bahia, and in Piauí, the state with the highest incidence, 100 cases have already been diagnosed. This mycosis has already been diagnosed in dogs and armadillos (Dasypusnovemcinctus), and *C. immitis* has been taken from soil samples collected from armadillo burrows in the states of Piauí and Ceará. It is likely that many cases are misdiagnosed as non-specific pneumonia, tuberculosis or even pneumoconiosis/silicosis.

Cases in north-eastern Brazil have a bimodal zonality, occurring more consistently at the start of the rains in January and in the hot, dry months in the region: September (16.7%); October (10.0%); and November (20.0%). In particular, January was the month with the highest incidence of the disease (30%) in the state of Piauí, coinciding with the start of the rains and the highest rainfall throughout the state.

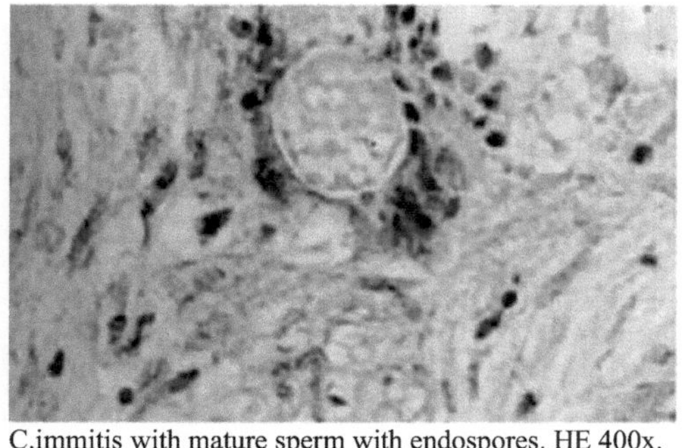

C.immitis with mature sperm with endospores. HE 400x.
Source: Revista da Sociedade Brasileira de Medicina Tropical 31:559-562, Nov-Dec, 1998.

TRANSMISSION

By inhalation of arthroconids (the form of the fungus in the soil). Transmission by inoculation, especially as a result of laboratory accidents, is relatively common. Transmission during pregnancy is rare and, when it does occur, neonatal death can occur.

CLINICAL FORMS AND SYMPTOMS

There are three main clinical forms of coccidioidomycosis: primary pulmonary, progressive pulmonary and disseminated pulmonary.

Primary pulmonary coccidioidomycosis:

Coccidioidomycosis is characterised by pulmonary manifestations that generally appear one to three weeks after exposure to the fungus. Around 60% of infected individuals progress to a spontaneous cure without showing any clinical or radiological manifestations. The remaining 40% usually show manifestations of acute respiratory illness, simulating flu, with fever, night sweats, cough and/or pleuritic chest pain.

The manifestations appear between 10 and 15 days after exposure to the fungus and the intensity of the symptoms depends directly on the infectious load, ranging from a flu-like state to a severe non-specific respiratory infection, with high fever, chest pain, cough with or without sputum, accompanied by general symptoms or allergic manifestations, particularly erythema nodosum.

This form of coccidioidomycosis usually regresses spontaneously to a cure within 30-60 days, even without antifungal treatment. However, around 5% of these patients develop residual lung lesions, usually solitary nodules, which in most patients are asymptomatic. These cases are often diagnosed after surgical removal on suspicion of lung carcinoma.

However, 5% of these patients still develop thin-walled, solitary, juxtapleural cavities, which can regress spontaneously in around 2 years. In some cases, especially in diabetic or immunocompromised patients, the acute pulmonary form does not regress, evolves into chronic pneumonia and is characterised by the formation of lung cavities.

The lungs can be affected diffusely, as a result of inhaling a large quantity of infective arthroconidia or as a late and secondary presentation resulting from haematogenous dissemination. These forms present with multiple diffuse infiltrates, the largest of which may present cavities that cause severe respiratory manifestations and can lead to respiratory failure. The evolution can also be fulminant, with septic shock.

Progressive pulmonary coccidioidomycosis:

Usually chronic, it evolves from primoinfection whose symptoms have not regressed after 2 months. It can present as: 1) nodular or cavitary lesions, sometimes representing a casual radiological finding; 2) fibrocavitary lung disease; 3) pulmonary miliary dissemination, with non-specific clinical and radiological manifestations. Due to its chronic progressive evolution, it is an important differential diagnosis with pulmonary disease.

Disseminated coccidioidomycosis:

Approximately 0.2% of patients with the primary pulmonary form develop dissemination of the lesions, predominantly to the skin, central nervous system and osteoarticular system. The presence of mediastinal or paratracheal lymphadenomegaly is indicative of dissemination. The disseminated form usually evolves acutely, affecting several organs or systems, and is rapidly fatal if not diagnosed and treated in time.

However, it can progress in a protracted manner, spreading to various organs, with periods of remission and recrudescence, regardless of antifungal treatment. The most frequently disseminated lesions are found on the skin, in the central nervous system, in bones and joints and in the genitourinary system.

Skin lesions are the most common extrapulmonary localisation, with a predilection for the face, usually papular or verrucous, but plaque forms, superficial abscesses, pustules and granulomatous lesions can also appear.

DIAGNOSIS

Clinical Diagnosis

The major issue in the diagnosis of coccidioidomycosis is the fact that the multidisciplinary health team does not know the fungi associated with this pathology and the lack of support from this team for laboratory diagnosis mediated by mycological diagnosis.

However, the clinical picture of this pathology must be differentiated from other infections, such as acute respiratory syndromes, tuberculosis and other mycoses that affect the lungs. However, it is necessary to study the epidemiology of the pathology in question in order to help differentiate between cryptococcosis and other mycoses that affect the whole body.

Laboratory diagnosis

For laboratory diagnosis to be effective, it is necessary to use biological samples to be analysed, such as: sputum, exudate from lesions, pus material, urine aliquots, aspirates from bone lesions, samples of cerebrospinal fluid and other relevant sites. In order for the samples to be analysed, 10% potassium hydroxide solutions must be prepared to observe the structures of *C.immitis*.

However, the observation of mature spheres with the presence of endospores is essential to define an accurate diagnosis so that the analyser does not confuse them with other fungi that have the same characteristic. However, in liquid samples, centrifugation should be carried out for up to 3 hours using potassium hydroxide.

Knowing the potential risk of these fungal samples, they are not cultured. However, if it is done, it should be done under the protection of a biological safety cabinet or hood. However, other ways of diagnosing this pathology are already being used, such as molecular biology techniques: PCR and exoantigen tests to visualise suspicious DNA structures.

Immunological tests are also of great value for diagnosis, including precipitation, complement fixation and immunodiffusion tests. The first test evaluates the precipitation of IgM-type antibodies from the acute phase of the disease. The second test studies the evidence of IgG antibodies in the more chronic forms of the disease. However, immunodiffusion tests have the same evaluative properties as the complement fixation test and are more specific for *C.immitis*.

Other tests used to detect these fungal agents are histcpathological examinations, which are obtained through biopsy samples of lesions caused by the fungus and suspicious necrotic material. However, another great solution for diagnosis are skin tests to visualise hypersensitivity, which are highly sensitive and specific.

PREVENTION

There is still no fully effective vaccine for treating coccidioidomycosis. However, this disease is not contagious and there is no need to control symptomatic patients.

TREATMENT

The specific drugs for the clinical stages of Coccidioidomycosis are fluconazole-based antifungals at dosages of 400 to 1200mg/day, intraconazole for the disseminated stages of the disease, amphotericin B for the most severe forms and voriconazole, which is a new drug being tested with a broad spectrum of action. If patients do not respond to antifungal drugs, surgery is necessary to resect the lesion.

However, in patients who are HIV-1 victims it is necessary to carry out therapy when the CD+ cell count is greater than 250 cells/^l. However, if the CD4+ cell count rises, treatment should be stopped immediately. However, if the patient is suspected of having menigitis, the medication should be given for life.

PARACOCCIDIOIDOMYCOSIS

INTRODUCTION

Paracoccidioidomycosis is defined as a systemic mycosis that mainly affects men who work in the fields and is considered to be the most important fungal infection in Latin America and Brazil. Its symptoms are mainly associated with the lungs, due to inhalation of the fungus, which can spread to various systems, causing second-level lesions that affect mucous membranes, lymph nodes, skin and glands. It became a subject of great interest to the medical community after the first papers published by Adolpho Lutz, which brought to light data on research, frequency, diversity of clinical manifestations and therapeutic challenges. The last major work on PCM dates back 20 years by FRANCO et al. Therefore, new work that adds to the knowledge obtained in recent years is of great scientific value and necessity.

HISTORY

This mycosis was first described by Adolhpo Lutz in 1908 and a few years later Afonso Splendore was the one who directed experiments on the etiological agent of the mycosis and classified it in the genus Zymonema brasiliensis and designated it as paracoccidioidomycosis in 1974 by mycologists at a meeting in Medellín (Colombia) and 15 years after this meeting, the clinical forms of the disease were classified at the III International Meeting on Paracoccidioidomycosis.

ETIOLOGICAL AGENT AND VIRULENCE FACTORS

Paracoccidioidomycosis is a systemic mycosis with granular characteristics caused by a fungus called Paracoccidioides brasiliensis, which grows in tissues at temperatures between 35°-37°C. It can be observed microscopically as soft, rough colonies, which are formed microscopically by oval or elongated cells with multiple buds. When grown at 13-23°C, this fungus develops mycelial morphology, characterised by white, small and irregular colonies.

The ecological niche of P. brasiliensis is currently unknown and associated with the fact that the fungus remains in symbiosis with man for a long time, this significantly hinders the identification of its ecology, so it is believed that it is found in saprophytic form in nature and in fungal reservoirs such as dog food and penguin faeces.

The cell wall of P. brasiliensis is made up of layers of lipids and proteins, polysaccharides a-(1,3)-glucan and P-(1,3)-glucan and the extracellular antigens gp43, gp70 and gp30, all examples of important virulence factors. In addition, the fungus has a glycoprotein gp43, which binds to laminin and inhibits MHC-II expression and macrophage function, which is also presented as Ag. It also has a high dimorphism.

CLINICAL MANIFESTATIONS AND SYMPTOMS

The host is probably infected by inhaling hyphae or conidia. These penetrate the upper airways and initially settle in the lungs. (WANKE and AIDÊ, 2009) It reaches the alveoli, multiplies and forms a primary focus, spreads to the regional lymph node and this constitutes the primary complex. From there, it can spread, causing lesions in any of the hosts, called a metastatic focus, and may or may not cause symptoms or apparent signs. When a person is affected by the asymptomatic form, there are three types of progression:1 CP regresses with the destruction of the fungi; it regresses with the persistence of the fungi or progresses to the development of the disease.

When PCM establishes itself as a disease, it can be acute or chronic. The former is prevalent in children and adolescents of both sexes and accounts for around 5% of cases, characterised by skin lesions, lymph node involvement of the cervical, thoracic and abdominal chains, profound intestinal changes and hepatosplenomegaly. The chronic type manifests mainly in adults and is responsible for 90% of cases. This form mainly affects the lungs, characterising the chronic form.

DIAGNOSIS

In the acute form, PCM has non-specific signs and symptoms. When assessing patients with the chronic form, the examination should include signs and symptoms related to integumentary, pulmonary, laryngeal, lymphatic, adrenal and central nervous system involvement. The tests to be carried out in the acute and chronic phases as a way of investigating are: simple chest X-ray, abdominal ultrasound, full blood count, erythrocyte sedimentation rate, liver function tests, urea, creatinine, sodium and potassium and confirmation of PCM is obtained by visualising the fungus in clinical specimens or tissue biopsy.

TREATMENT

Due to the polymorphism of the lesions, diagnosis and treatment are difficult, resulting in progression of the disease and potentially leaving disabling sequelae. Treatment includes the use of antifungal drugs associated with nutritional balance and treatment of present sequelae and co-morbidity. Among the antifungal drugs are ointraconazole and amphotericin B.

TINEA

Also known as ringworm or more popularly as chilblains, they are superficial skin infections caused by filamentous fungi of the genera *Microsporum, Epidermophyton and Trichophyton* - collectively known as Dermatophytes - which are able to use keratin as a source of nutrients and are among the most common diseases in the world, causing significant chronic morbidity.

It can occur in any area of the body, but is most common in folds such as the armpits, groin, between the fingers and toes. Symptoms appear between 4 and 10 days after contact, and are common:

S Appearance of white or reddish spots that cause itching;

S Edge of lesion evident, crusting in some cases;

S Pain and cracking (if the affected area is the feet);

S Skin eruptions in the shape of rings or circles with slightly raised edges even if the skin in the middle of these eruptions appears healthy;

As a result, this type of infection can spread rapidly throughout the body and, depending on the area affected, can be given different names.

Tinea capitis:

Infection of the scalp is called *Tinea capitis* and is a disease that can produce lesions ranging from hair thinning to definitive alopecia. This is almost exclusively a childhood disease, mainly of school-age children. However, adults rarely get Tinea *capitis*, although they can develop *Tinea corporis* (of the head or neck) through spreading it on a child's scalp.

The infection is often asymptomatic and goes unnoticed, consequently spreading to close contacts, constituting a major public health problem. Due to its flaking, it looks similar to dandruff and can initially be masked with hair oils.

Tinea unguium (Onychomycosis)

Tinea unguium is a type of nail infection caused by dermatophytes, also known as onychomycosis, and is commonly caused by fungi of the Trichophyton and Epidermophyton genera. In general, the nail plates show discolouration and can become thick, brittle or distorted, suffering from onycholysis (elevation of the nail plate). Onychomycosis of the toenails is more common than the disease of the fingernails.

However, feet are often more exposed to damp places, not only when walking barefoot in public places, but also when spending much of the day inside socks and shoes. On hot days, bare feet can spend several hours covered and damp from sweat. Thus, heat, lack of light and humidity are all that a fungus wants in order to proliferate.

In addition, the toes are the furthest part of the body from the heart and are not as well vascularised as the fingers. As a result, the body's antibodies and defence cells don't reach the toenails as easily as they do other parts of the body. The infection affects around 10 per cent of the adult population and 20 per cent of the elderly.

Contact with the fungus alone is not usually enough to acquire the infection. There usually need to be small lesions between the nail and the skin for the fungus to penetrate under the nail and take up residence. It is also necessary for the nail to be frequently exposed to humid environments so that the fungus can multiply more easily.

The disease is usually atypical and aggressive in patients with untreated HIV infection. However, immunosuppressed patients may be predisposed to a more severe and frequent infection, as well as a failure to respond to conventional doses of antifungal drugs.

Patients with psoriasis are at increased risk because the abnormal psoriatic nail represents a gateway for fungi. As the two conditions can be clinically similar, laboratory confirmation is necessary before starting oral treatment and should be repeated at the end of treatment.

Tinea Corporis (tinea on the body):

Tinea corporis, also known as tinea do corpo - in Brazil it is popularly known as impingem - is a skin infection that manifests itself on various parts of the body, such as the face, neck or feet. This type of infection can be acute (sudden onset and rapid spread) or chronic (slow extension of a slightly inflamed rash).

Acute *Tinea corporis* causes red patches, itching and inflammation, and can also lead to the formation of pustules. Often the cause is a fungal infection that affects animals (zoophilic) while chronic *Tinea corporis* tends to affect the folds of the body (spread of *Tinea cruris*).

However, if the disease spreads, it is more difficult to treat and relapses are more likely. This is probably due to the decrease in the skin's natural resistance to fungi. Nevertheless, tinea on the body typically manifests as one or multiple ring-shaped scaly lesions with central attenuation and a slightly raised, erythematous border. The border may show pustules or follicular papules and the presence of itching is variable.

Tinea Cruris (Inguinalis):

Caused by the fungi *T. rubrum, T. mentagrophytes and E. floccosum,* it is a very common infection, especially in summer, when sweating and heat make it easier to develop, maintain or relapse; it is more common in adult males, but recent studies show that women who have just come out of puberty, who are overweight or who wear very tight trousers are also prone to developing the infection.

The lesions invariably affect both thighs and can progress to the perineal, gluteal and abdominal regions. They are characterised by erythema and scales, with demarcated borders, in which there are small vesicles that may be exudative.

In this active phase, itching is constant. During the quiescent periods, the lesions become darkened and the itching subsides. In the evolution, which is usually chronic, periods of activity alternate with periods of calm

Tinea pedis and Tinea manuum (Tinea of the feet and hands):

It is the most common dermatophytosis of all, and in summer, due to the heat and humidity, there is a possibility of worsening or recurrences. In addition to humidity, diabetes and HIV increase the rates of this type of infection.

It is more common in adults, although there has been an increase in childhood in recent years, probably due to the use of trainers and thick, synthetic socks that increase sweating and maceration of the feet. In our country, the fungi most commonly involved are *T. rubrum, T. mentagrophytes and E. floccosum.*

The most common localisations are the soles of the feet and the interdigital spaces of the last three toes, while the nail folds and the backs of the feet are less affected. Thus, according to the clinical picture, it is variable and can be subdivided into four forms   :

• Intertriginous - chronic form with maceration and desquamation of the interdigital spaces, with possible itching and cracking; - Vesicular - subacute form of tinea on the feet. Vesicles or vesicopustules affect localised areas of the interdigital spaces and may suffer secondary infection; - Scaly - scaly or erythematous-scaly lesions, with frequent itching and a discreet inflammatory reaction;

• Margined - scaling with a circled appearance at the edges.

Tinea on the hands is much rarer than tinea on the feet and is caused by the same species of fungus. The back of the hands is affected more often than the back of the feet.

Tinea barbae (Tinea of the beard):

It is relatively rare in our country and is caused by fungi of the Trichophyton genus. It affects hirsute men and women. It can usually lead to scaling, follicular pustules and erythema. There may also be alopecia in the centre of the lesions.

It comes in three forms:

a) inflammatory type, in which the lesions are inflammatory and suppurative;

b) herpes circinatus type, consisting of annular lesions (erythematous-papular-vesicular-scaly at the edges), whose growth is centrifugal;

c) sicosiform type, which resembles bacterial folliculitis.

REFERENCES

AIDÉ M A. Article Chapter 4 - Histoplasmosis. J. bras.pneumol. vol.35 no.11 São Paulo Nov. 2009.

ALBORNOZ, M. B. Paracoccidioidomycosis - Infection. In: DEL NEGRO, G.; LACAZ, C. S.; FIORILLO, A. M. Paracoccidioidomycosis. São Paulo: Sarvier - EDUSP, chap. 8, p. 91-96, 1982.
ALVARES, A.C., SVIDZINSKI, E.I.T., CONSOLARO, L. E.M. Vulvo-vaginal cnadidiase: host predisposing factors and yeast virulence. J Bras Patol Med Lab. v.43. n.5. p.319-327. October 2007.

ALVES, A.I. CAMARGO, P. F. GOULART, S.L. PCR identfication and antifungal sensitivity of clinical vaginal isolates of Candida sp. Revista da Sociedade Brasileira Tropica 43(5):575-579, Sep-Oct, 2010.

AMATO, ACM; DICHTCHENIAN, RB; MORILLO, MG. Report of a case of pulmonary histoplasmosis. Ver. Fac.Ciênc.Méd. Sorocaba, v.4, n.1-2,p.62-65,2002.

ARAUJO, R; PINA-VAZ; C.; RODRIGUES, A. G. Surveillance of airborne Aspergillus in a Portuguese University Hospital,Mycoses, 2005. 48(2). pp.45.

ARENAS R. Dermatofitosisen México. Revista Iberoamericana de Micologia 19:63-67, 2002.

ASTE N, PAU M, BIGGIO P. Tinea capitis in adults. Mycoses 39:299-301, 1996.

AZULAY, R.D.; AZULAY, D.R. Dermatologia. 2ª ed. Guanabara Koogan, Rio de Janeiro, 1997.

BALS, R.Aspergillusfumigatus conidia induce interferon-b signalling in respiratory epithelial cells,European Respiratory Journal, 2012, 39, pp.411-418.

BARBEDO,S.L., SGARBI,B.G.D. Candidiasis. DST- J Bras Doenças Sex Transm 2010:22(1): 22-38 - ISSB: 0103-4065 - INSS online 2177-8264.

BHABHRA, R.; ASKEW, D.S. Thermotolerance and virulence of Aspergillusfumigatus: role of the fungal nucleolus,MedMycol, 2005, 43(1), pp. 87-93.

BIVANCO, FC; MACHADO, CDS; MARTIN, EL. Cutaneous cryptococcosis. Dermatology Service of the ABC Medical School. Arq.Med.ABC.2006; 31(2):102-9.

BRUMMER, E.; CASTANELA, E.; RESTREPO, A. Paracoccidioidomycosis: an update. Clinical Microbiology Reviews, v. 6, p. 89-117, 1993.

CAMARGO, C.K., ALVES, F.R.R., BAYLÃO, A.L., RIBEIRO, A.A., ARAUJO, S.A.L.N., TAVARES, N.B.S., SANTOS.R.H.S. Abnormal vaginal secretion: sensitivity, specificity and agreement between clinical and cytological diagnosis. Rev Bras Ginecol Obstet. 2015; 37(5):222-8.

CASAGRANDE, JLM; CRUZ, DVN; MACHADO, D S; ALMEIDA, M F O; OGATA, DC. Gastroduodenal ulcers as a clinical manifestation of disseminated histoplasmosis in a human immunodeficiency virus-positive patient: case report. Rev. Soc. Clin. Méd;9(6), Nov.-Dec. 2011.

CHANDRASEKAR, P. Management of invasive fungal infections: a role for polyenes,Journal Antimicrobial Chemotherapy, 2010, 66, pp. 457-465.

COLOMBO, L.A., GUIMARAES, T. Candiuria: a clinical and therapeutic approach. Revista da Sociedade Brasileira de Medicina Tropical 40(3):332-337, mai-jun, 2007.
Consensus on cryptococcosis - 2008. Journal of the Brazilian Society of Tropical Medicine. 41(5);524-544 Sep-Oct.2008.

CROCCO.I E, SOUZA, M.V., MIMICA, J.M.L, RIUZ.B.R.L., MURAMATU.H.L., ZAITZ, C., GARCIA, C. Identification of candida species and in vitro antifungal susceptibility: study of 100 patients with superficial candidiasis. An. Bras. Dermatol, Rio de Janeiro, 79(6):689-697, Nov/Dec. 2004.

CUCÊ, L.C.; SALEBIAN, A.; SAMPAIO, S.A.P. Systemic treatment of superficial mycoses with ketoconazole. An. Bras. Dermatol, 1982; 57:5.

DEMCHOK, J.P.; MELETIADIS, J.; ROILIDES, E.; WALSH, T.J. Comparative pharmacodynamic interaction analysis of triple combination of caspofungin and voriconazole or ravuconazole with subinhibitory concentrations ofamphotericin B against Aspergillusspp. Mycoses. 2009.

ENDO, S.; KOMORI, T.; RICCI, G. SANO, A.; YOKOYAMA, K.; OHORI, A.; KAMEI, K.; FRANCO, M.; MIYAJI, M.; NISHIMURA, K. Detection of gp43 of Paracoccidioidesbrasiliensis by the loop-mediated isothermal amplification (LAMP) method. FEMS Microbiology letters, Amsterdam, v. 234, n.1, p. 93-97, 2004.

FERRAZZA, H.S.H.M., MALUF, F.L.M., CONSOLLARO, L.E.M., SHINOBU, S.C., SVIDZINSKI, E.I.T., BATISTA.R.M. Caracterização de leveduras isoladas da vagina e sua associação com candidiase vulvo vaginal em duas cidades do sul do Brasil. Rev Bras Ginelocol Obstet. 2005;27(2):58-63.

FERREIRA, M S; BORGES, A S. Histoplasmosis. Revista da Sociedade Brasileira de Medicina Tropical 42(2):192-198, Mar-Apr, 2009.

FERREIRA, M. S.; FREITAS, L. H.; LACAZ, C. S.; DEL NEGRO, G. M. B.; MELO, N. T.; GARCIA, N. M.; ASSIS, C. M.; SALEBIAN, A.; HEINS-VACCARI, E. M. Isolation and characterisation of a Paracoccidioides brasiliensis starin from a dogfoodprobalycontaminatedwithsoil in Uberlândia , Brazil. Journal of Medical and Veterinary Mycology. Oxfordshine, v. 28, p. 253-256, 1990.

FILHO, A.D.; DEUS, B.C.A.; MINEZES, O.A.; SOARES, S.A.; LIRA, A.L.A. Cutaneous-mucosal manifestations of coccidiodomycosis: a study of thirty cases from the states of Piaui and Maranhão. Rev. Dermatol. Vol 35 no.1. Rio de Janeiro. Jan-Feb: 2010.

FILHO.D.A. Coccidioidomycosis. J.Bras.pneumol. vol.35 no.9 São Paulo, sep. 2009.
FRANÇA A, S; SILVA J, B. Diagnosis and treatment of cryptococcosis of the central nervous system. Arq. Neuro-psiquiat. São Paulo.vol.26, n°2. June-1968.

FRANCO, M.F.; MONTENEGRO, M.R.G. Pathological anatcmy. In: DEL NEGRO, G.; LACAZ, C. S.; FIORILLO, A. M. Paracoccidioidomycosis. São Paulo: Sarvier - EDUSP, 1982. chap. 9, p. 97-117.
FRANCO, M; LACAZ, C.S; RESTREPO, A; DEL NEGRO G. Paracoccidioidomycosis. Boca Raton: CRC Press, p 409. 1994.

GENTIL F, A; DIAS D, T; PACHECO D, F; LUZ E, A; COSTA L, G, F, S; BRUM R, M, O. Cryptococcosis: Case report. Acta Biomedica Brasiliensia.vol 7. N.2. December.2016.

GOLDANI, L. Z.; SUGAR, A. M. Paracoccidioidomycosis and AIDS: an overview. Clinical Infectious Diseases, v. 21, n. 5, p. 1275-1281, 1995.

GOMES, A P; ALBUQUERQUE, VS; RAMOS, OF; GUIMARÃES, PE; ANTÔNIO, VE. Natural history of histoplasmosis. J. bras.med; 95(1): 46-50, jul. 2008.

HOLANDA, R.A.A., FERNANDES, S.C.A., BEZERRA, M.C., FERREIRA, F.A.M., HOLANDA, R.R.M., HOLANDA, P.C.J., MILAN, P.E. Vulvovaginal candidiasis: symptoms, risk factors and concomitant anal colonisation. Rev Bras Ginecol Obstet. 2007; 29(1):3-9.

HOPE, W.W; WALSH, T.J.; DENNING, D.W. Laboratory diagnosis ofinvasiveaspergillosis, Lancet Infect Dis, 2005, 5, pp. 609-622.

KURUP, V.P.; KUMAR, A. Immunodiagnosis of aspergilosis.ClinMicrobiol Rev 1991;4:439-56.

LACAZ, C. S.; PORTO, E.; MARTINS, J. E. C.; HEINZ-VACCARI, E. M.; MELO, N. T. Tratado de micologia médica Lacaz. 9th edition, p. 649-713, 2002. São Paulo, SP: Sarvier.

LACAZ, C.S.: Mycoses. In: Veronesi, R.: Infectious and Parasitic Diseases. 6ª ed. Guanabara Koogan. Rio de Janeiro, 1976.

LACAZ, C.S.; PORTO, E.; MARTINS, J.E.C.: Mycologia Médica. 7th ed. Sarvier. São Paulo, 1984.

LONDERO, A. T. Paracoccidioidomycosis: pathogenesis, clinical forms, pulmonary manifestations and diagnosis. Brazilian Journal of Pneumology, v. 12, p. 41-60, 1986.

LORENZONI, P. J.; CHANG, M. R.; PANTAGO, A. M. M.; SALGADO, P. R. Paracoccidioidomycotic meningitis. Arquivos de Neuropsiquiatria, v. 60, n. 4, p. 10151018, 2002.

MARTINS E M L, MARCHIORI E, DAMATO S D, POZES A S. Acute pulmonary histoplasmosis: report of a microepidemic. RadiolBras2003;36(3): 147-151.

MARTINS, RC; ADDOR, G; MONTEIROA, S; FRANCO, CAB; NIGRI, D H. Pulmonary histoplasmosis in a private clinic in Rio de Janeiro. Pulmão RJ 14(3): 197201, 2005.
MINELLI, L. Superficial mycoses, sporotrichosis and mycetomas. In: Amato Neto, V.; Baldy, J.L.: Doenças Transmissíveis, 3a ed. Sarvier, pág. 623-634, São Paulo, 1989.

MINELLI, L; MINELLI, L.D.; MINELLI, H.H.: Superficial mycoses - How to diagnose and treat. Rev. Bras. Med., 1993; vol. 50, p. 903-918.

MORAES, MF; VITAL, LP. Meningitis due to Cryptococcus neoformans- Diagnosis and treatment. Centro Universitario Faculdade Metropolitanas unidas. São Paulo-SP.

MORAES, P.A.M.; MARTINS, M.L.R.; LEAL, R.I.I.; ROCHA, S.I.; JUNIOR, M.P. Coccidiodomycosis: a new Brazilian case. Revista da Sociedade Brasileira de Medicina Tropical 31(6):559-562, Nov-dez, 1998.

OLIVEIRA, F; BAZAN, C; SOLIVA, A; RITZ, R; FAGUNDES, E; CAMARGO, G; AUGUSTO, M. Cryptococcosis- Revista Cientifica eletrônica de Medicina Veterinaria. July 2008. São Paulo-sp.

PANTOJA, MGL; SILVEIRA, DMR; SILVA, LD. Disseminated cryptococcosis in immunocompetent patients: Case report. Federal University of Pará (UFPA).

PEDROSO, RS; CANDIDO, RC. Laboratory diagnosis of cryptococcosis. University of São Paulo (USP). Newslab.ediçao77. 2006.

PONTES, F. S. C.; PONTES, H. A. R, MOREIRA, C. R., DANIN, G. A.; PEREIRA, E. M. Paracoccidioidomycosis: general aspects and case report. Brazilian Journal of Surgery and Implantology. V. 7, p. 73-77, 2000.

QUEIROZ, JPAF; SOUSA, FDN; LAGE, RA; IZAEL, MA; SANTOS, AG. Cryptococcosis - A Bibliographical Review. Acta veterinária brasílica. V.2, n.2, p.32-38- 2008.

RODRIGUES, T.M., GONÇALVES, C.A., ALVIM, T.C.M., FILHO, C.S.D., ZIMMERMMANNN, B.J., SILVA,L.V., DINIZ, G.C. Association between vaginal secretion culture, sociodemographic characteristics and clinical manifestations of patients diagnosed with vulvovaginal candidiasis. Rev Bras Ginecol Obstet. 2013;35(12):554-61.

ROSSINI T F, GOULART L S. CLASSICAL HISTOPLASMOSIS: REVIEW. RBAC, vol. 38(4): 275-279, 2006.

SAMBATAKOU, H; DUPONT, B; LODE, H; DENNING, DW. Voriconazole treatment for subacute invasive and chronic pulmonary aspergillosis. Am J Med. 2006;119(6):527.e17-24.

SHIKANAI-YASUDA, M. A.; TELLES FILHO, F. Q.; MENDES, R. P.; COLOMBO, A. L.; MORETTI, M. L. Guideliness in paracoccidioidomycosis. Revista as Sociedade Brasileira de Medicina Tropical, v. 39, n. 3, p. 297-230, 2006.

TARANTINO, A. B.; GONÇALVES, A. J. R.; CAPONE, D.; AIDÊ, M. A.; LAZERA, M. S.; WANKE, B. Pulmonary Mycoses. In: TARANTINO, A. B., editor. Pulmonary Diseases. Rio de Janeiro: Guanabara Koogan, 2002, p. 416-450.

TOGASHI, H.R.; AGUIAR, B.M.F.; FERREIRA, B.D.; MOURA, M.C.; SALES, M.T.M.; RIOS, X.M. Pulmonary and extrapulmonary coccidioidomycosis: three cases of incidence in the interior of Ceara. Brazilian Journal of Pneumology. Volume 35- num 3. March - 2009.

UNIS G, PÊGAS K L, SEVERO L C. Pulmonary histoplasma in Rio Grande do Sul. Revista da Sociedade Brasileira de Medicina Tropical 38(1): 11-14, Jan-Feb, 2005.

UNIS G, SILVA V B, SEVERO L C. Disseminated histoplasmosis and AIDS. Importance of the culture medium for the clinical-bronchoscopic specimen. Revista da Sociedade Brasileira de Medicina Tropical 37(3)234-237, mai-jun, 2004.

VALLE, A. C. F., COSTA, R. L. B. Paracoccidioidomycosis. In: BATISTA, K. S., IGREJA, R. P.; GOMES, A. D.; HUGGINS, D. W. Medicina Tropical: abordagem actual das doenças infecciosas e parasitárias. Rio de Janeiro: cultura Médica, 2001, p. 943-958.

VICENTINI A P, PASSOS A N, SILVA D F, BARRETO L C, ASSIS C M, FREITAS R S. Article Histoplasmosis: an occupational risk among fieldwork researchers? RevInst Adolfo Lutz. 2012; 71(4):747-52.

ZAMBONI M; ROMANO S; TOSCANO E. Histoplasmosecavitary. Pulmão RJ; 13(1): 54-57, jan.-mar. 2004.

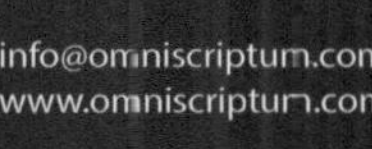

yes I want morebooks!

Buy your books fast and straightforward online - at one of world's fastest growing online book stores! Env ronmentally sound due to Print-on-Demand technologies.

Buy your books online at
www.morebooks.shop

Kaufen Sie Ihre Bücher schnell und unkompliziert online – auf einer der am schnellsten wachsenden Buchhandelsplattformen weltweit! Dank Print-On-Demand umwelt- und ressourcenschonend produziert.

Bücher schneller online kaufen
www.morebooks.shop

info@omniscriptum.com
www.omniscriptum.com

OMNIScriptum